Relativity in Illustrations

Jacob T. Schwartz

Professor of Mathematics and Computer Science
New York University

Designed and Illustrated by
Felix Cooper

DOVER PUBLICATIONS, INC., New York

Published in Canada by General Publishing Company, Ltd., 30 Lesmill Road, Don Mills, Toronto, Ontario.
Published in the United Kingdom by Constable and Company, Ltd.

This Dover edition, first published in 1989, is an unabridged republication of the work originally published in 1962 by New York University Press. The elements of the illustrations originally in color appear in black-and-white in the present edition, and corresponding changes have been made in the text.

Manufactured in the United States of America
Dover Publications, Inc., 31 East 2nd Street, Mineola, N.Y. 11501

Library of Congress Cataloging-in-Publication Data

Schwartz, Jacob T.
  Relativity in illustrations.

  Reprint.
  Bibliography: p.
  1. Relativity—Popular works.   I. Title.
QC173.57.S39   1989   530.1'1   88-33429
ISBN 0-486-25965-X

CONTENTS

PREFACE

Relativity theory has for many years been the object of exceptionally widespread curiosity. I hope that the present small book will help inform my intelligent fellow citizens, in every occupation and of every degree of education, who, intrigued by rumors of the surprising things this theory has to say, have wondered what all the talk is about. I shall be particularly gratified if this book finds favor with that audience of which I have been particularly conscious during its composition—the clever young people who, sometime between the age of thirteen and nineteen, wake up, as did Einstein himself, to their calling as scientists.

It is my final pleasure to thank those friends who helped in the various stages of preparation of this book: Beatrice Bookchin, who suggested that it be written; Virginia Davis, for reading the manuscript and suggesting many improvements; Steve Seltzer, for his criticisms, based, as always, on a scrupulously honest determination to get to the bottom of things; Adrienne Winogrand, for imparting (I hope) a little of her charm; my wife Sandra, for her many clarifications; Mr. Benjamin Bold and his students at Seward Park High School, for valuable service as test readers;

Marc Drogin and Carl Bass for their help with the early stages of the artwork; my colleagues Louis Nirenberg, Lipman Bers, and Anneli Lax, for continuing encouragement; Ursula Burger and Ruth Murray, for invaluable assistance in preparation of the manuscript.

Mr. Felix Cooper's book design has more than fulfilled my aesthetic hopes. To all of them, and to others unmentioned, I extend my gratitude.

Jack Schwartz

Malaga, Spain

January 1962

Relativity in Illustrations

INTRODUCTION

The first of the revolutionary new physical theories
developed in the present century was Einstein's famous theory of relativity. As we shall see, this theory
overthrew many notions which previously had
seemed utterly unquestionable: basic attitudes
toward time and space which went back thousands of
years. Now, even though Einstein's ideas are pro-
found, they are not complicated: perhaps their greatest beauty is their crystalline simplicity.

It is the object of the present small book to present
these ideas. We shall begin with very general con-
siderations; then, gradually introducing the ideas
from ordinary geometry needed to follow the develop-
ment of Einstein's ideas, pass over to more specific
notions.

The reader should, above all, be reflective and
careful to master each idea as it occurs. Even where
few words are used, much that is essential may be
said.

WHAT IS TIME?

The question at first seems foolish, because we are so sure we know. However, since Einstein was able to make such interesting discoveries by asking this question seriously and by answering it carefully, we ask again:

WHAT IS TIME?

and also,

WHAT IS SPACE?

Let us study the first question first.

WHAT IS TIME?

We think we know, because we seem to feel time passing constantly. Time, we feel at first, is that which passes; that whose passage separates the earlier from the later. What does this mean? It means that our experiences are related to each other as earlier and later—that some things happen first, and others happen afterward—that when the later things happen, we can mostly remember the earlier things that have happened, but that when the earlier things happened, we could not remember the later things which were to happen, but could only guess them.

First we are little and go to school. Then we graduate. Then we work and marry. Then we have children. First they are little and stay in the house. Then they go to school. Then they are big and go away. Then we are old.

Time is like a wire, and we are like beads being pushed along the wire, from earlier to later, without any return. This is what we can feel directly. But not more than this.

If we want to know more about time than what is earlier and what is later; if we want to know how much earlier and how much later, we can no longer rely on our feelings, on our direct perceptions. Our direct perception of time is merely qualitative. Some days seem long, others short. When we are children, hours seem very long, and the years between birthdays seem to be ages. Later, days, weeks, and years seem to vanish in a moment. To understand time, not merely as a qualitative *after* and *before*, but as a quantitative *THAT MUCH after* and *THIS MUCH before*, we must make use of our physical experience.

Night follows day, and day night; and in each night there is one instant when we see the last star in the Big Dipper at its highest. The intervals between these instants we call a "day." The pendulum swings from right to left, and from left again to right, and we call the intervals between the instants when we see the pendulum at its highest "seconds." The tiny regulator spring in our watch ticks in and out, in and out, and drives the hands of the watch round and round, over and past the marks in the watch face. By adjusting the speed of our watches we can make the hand pass over the successive marks on the watch face at exactly the same instants when the pendulum is at its highest. Then the intervals between the instants when the hand of the watch is over a mark on the watch face are also seconds.

We see from all this how our quantitative idea of time is taken from our physical experience. We arrange our notion of equal intervals of time in such a way as to be able to say of certain simple repetitive physical processes that they repeat themselves in equal intervals of time. When we have arranged our notion of equal intervals of time in this way, we find that many physical happenings have a simple description. The last star in the Big Dipper is at its highest once every twenty-four hours. The pendulum goes from side to side once each second. The flywheel in an engine revolves eight hundred times a minute. A radio wave oscillates seven million times a second, another radio wave oscillates eight million times a second. From the fact that we have arranged our notion of equal intervals of time in such a way that so very many different physical happenings all have

a simple description, we know that we have successfully chosen our notion of equal intervals of time in
a way appropriate for the understanding of our physical world.

We must remember, however, that this success, like
all successes, can turn out to be short of absolute.
Above all, we must remember that our quantitative
notions of time *come from our physical experience,
can be made definite ONLY by reference to physical
experience, and are SUBJECT TO CHANGE if a reconsideration of the details of our physical experience
seems to warrant change.*

WHAT IS SPACE?

We have a direct qualitative perception of space
also. We see things by moving our eyes and head
left or right, up or down. A given object, when looked
at, may appear bigger, which we learn to call nearer,
or smaller, which we learn to call farther. To this extent, space is seen. In early infancy we learn to move
our hands while watching them, and find that certain
muscular adjustments bring our hands up or down,
left or right, nearer and farther. The fact that things
which are on the left "for looking" are also on the
left "for reaching," and that things which are nearer
"for reaching" are also nearer "for looking," gives us
confidence in our space perceptions.

But just as with time, so also, if we wish to arrive
at a quantitative notion of space, we must make use
of our physical experience, specifically experience
with measuring tapes, rulers, with calipers, microm-
eters, surveyors' transits, magnifying glasses and
microscopes, telescopes, etc., of the experience of fitting things together, and finding that sometimes pieces
are too big to fit, and that sometimes pieces are too
small to reach from one point to another no matter
which way you turn them.

In these ways, through looking, reaching, fitting,
and measuring, we develop quantitative notions of
space. From the fact that our quantitative notions of

space and our quantitative notions of time fit together in such a way that many physical happenings have a simple description, we know that we have successfully chosen our notions of space and time in a way appropriate for the understanding of the physical world.

We must remember that this success can turn out to be short of absolute. Our quantitative notions of time and space come from our physical experience, can only be made definite by reference to physical experience, and are subject to change if a reconsideration of the details of our physical experience seems to warrant a change.

Now it is time to pass from generalities to details. We want to look at the specific way in which certain events take place in time and space: the when and where of particular happenings. How can this be done conveniently in the pages of a book? [Remark to the mathematically trained reader: the method which we shall use is simply that of graphing the position in space of these events against time. The next few pages are intended to explain the details of this technique. Naturally, the points at issue will seem quite simple to you.] The best way to begin is to use a sort of comic-strip technique, giving a series of pictures of the scene we want to study at successive instants.

Beginning with a series of cartoons showing various typical sequences of events to be studied, we shall, in the next few pages, introduce a number of conventions which will enable us to simplify these cartoons more and more, thereby depicting the events in a more and more convenient way. We will ultimately come to a scheme for representing a sequence of events by a highly simplified sort of chart or diagram, which will have the great advantage of showing everything relevant for our later analysis and of ignoring all that is irrelevant. These charts will be as basic to the reader of what is to follow as blueprints are to the engineer.

We want to look at the specific and detailed order in which certain events take place in time and space.

Consider the showdown in the main street of Snake City between Dead-Eye Dick and Piute Pete as it so often appears (Dick on the left, Pete on the right).

BEGINNING — — —

Fig. 1

Fig. 2

— — — END

How did this happen? (X marks are bullet holes.) Well, say the witnesses, Pete drew his gun, shot at Dick; Dick drew his gun, shot at Pete. Each was hit and killed by the bullet of the other. But who shot first? Or did they shoot at the same time?

Did Dick shoot first, and Pete a little later, but before being hit by Dick's bullet?

NOON

Fig. 3

ONE SECOND LATER

Fig. 4

TWO SECONDS LATER

Fig. 5

THREE SECONDS LATER

Fig. 6

FOUR SECONDS LATER

Fig. 7

FIVE SECONDS LATER

Fig. 8

Did Pete shoot first, and Dick later, but before being
hit by Pete's bullet?

NOON

Fig. 9

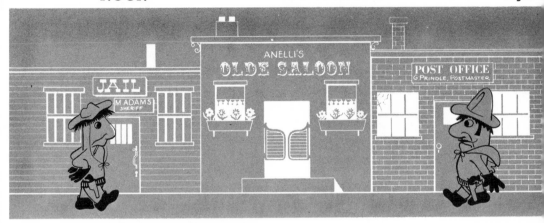

ONE SECOND LATER

Fig. 10

TWO SECONDS LATER

Fig. 11

THREE SECONDS LATER

Fig. 12

FOUR SECONDS LATER

Fig. 13

FIVE SECONDS LATER

Fig. 14

Did Dick and Pete shoot at the same time?

NOON

Fig. 15

ONE SECOND LATER

Fig. 16

TWO SECONDS LATER

Fig. 17

THREE SECONDS LATER Fig. 18

FOUR SECONDS LATER Fig. 19

FIVE SECONDS LATER Fig. 20

One can determine which of these possibilties is correct only by careful observation and measurement. The little drama of Pete and Dick is made up of a large number of separate events. Some of these events are: Dick fires his gun; Pete fires his gun; Dick is hit by a bullet; Pete is hit by a bullet; but also: Dick's bullet reaches the halfway mark (front door of saloon) in its flight to Pete; Pete's bullet reaches the three-quarter mark (left saloon window) in its flight to Dick, etc. Which of these events takes place first, which afterward, and which at the same time as a given event can be determined only by careful observation and measurement.

Since we shall have to study quite a number of cases like the case of Pete and Dick, we shall need a system for recording the results of our observations and measurements which is less clumsy than that of drawing a large number of detailed pictures of the scene at successive instants (these pictures, taken together, would clearly show the whole story of a given action).

For this reason, we shall make a number of simplifications in the way in which we draw our pictures.

First of all, we shall not bother to draw the background but shall merely indicate the position of important places by lines.

Fig. 21

Second, since all the action takes place along the length of the main street of Snake City, we shall not bother to show how far off the ground anything is, but shall show only where it is to the right or left along Main Street. This means that instead of drawing

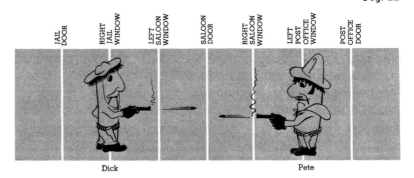

Fig. 22

Dick   Pete

we shall only draw

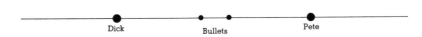

Dick   Bullets   Pete

When we use these simpler drawings, as we shall always do from now on, we deliberately forget that events, besides happening a certain distance to the right or left along Main Street, also happen a certain distance aboveground and also a certain number of feet closer to the foreground or farther away in the background. *That is, we deliberately forget that space is actually three-dimensional, and, for the sake of simplicity in description and pictorial representation, pretend that it is ONE-DIMENSIONAL.* All our events consequently take place on a "street," or along a "road" or "railroad track," so and so many feet to the left or right. Nothing important is lost by this simplification, but much is gained. The reader, however, *must be careful to remember that this simplification is being assumed*, or he may become confused.

With this simpler way of drawing, the whole set of pictures given on page 17 appears as

Fig. 23

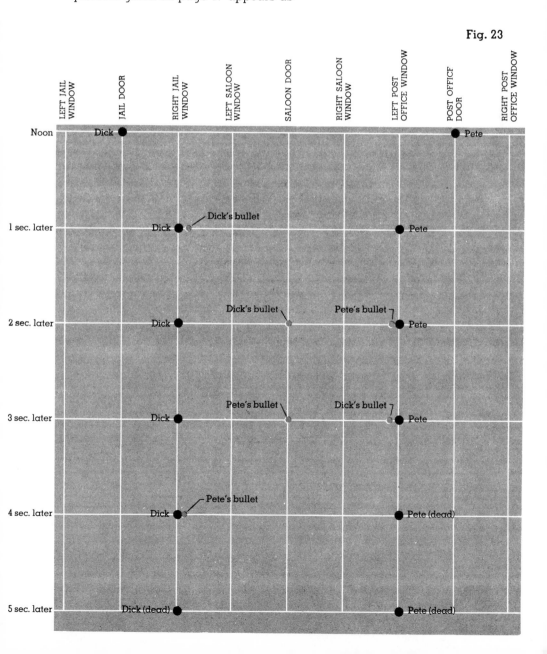

Our simpler way of drawing gives us room to describe the situation not only as it is after whole seconds have elapsed but also as it is after fractions of a second have elapsed. When this is done, many of the labels on Figure 23 become unnecessary.

Since ability to read a chart like Figure 24 will be essential in following the later sections of this book, it is well to pause and comment on this chart.

The first thing to realize is that a given horizontal line represents a given instant in time (one picture in

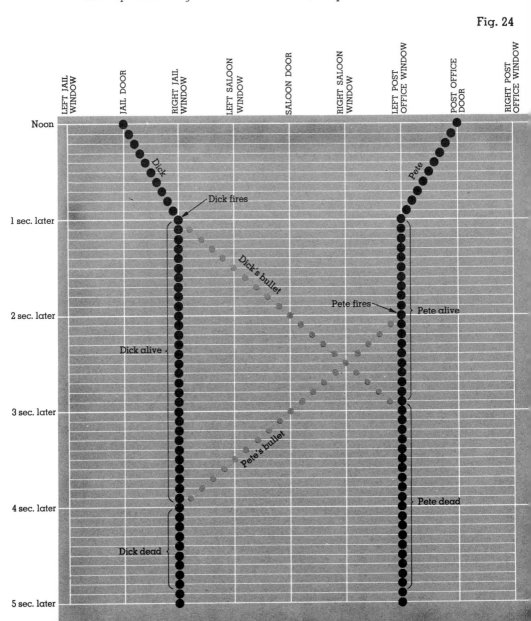

Fig. 24

a comic strip, one frame in a reel of motion-picture film), so that the dots on a given horizontal line represent the position of a number of objects at a given instant. For instance, Figure 24 shows that at 2.4 seconds after noon Dick is standing in front of the right jail window; Dick's bullet is a bit to the left of the right saloon window; Pete's bullet is a bit to the right of this same window; and Pete is standing in front of the left post-office window. Figure 24 shows the same sequence of events shown in Figures 23 and 3–8; it is less detailed than Figures 3–8 in that it shows less of the background, etc., but more detailed in that it shows the situation not only after 1, 2, 3, etc. seconds, but also after 1.1, 1.2, 1.3...seconds. That is, Figure 24 is able, by omitting those details which we wish to ignore, to show the details which are important to us—the detailed order in which certain events take place in time and space.

A number of features of the chart in Figure 24 are worth emphasizing. A period of time in which an object is not moving is represented in this chart by a vertical sequence of dots. Thus, for instance, from 1 second after noon on, Dick is continually in front of the right jail window; the dots representing his position at successive intervals of 1/10 second occur one under the other, all on the vertical line upon which all events occurring in front of the right jail window are recorded. These dots consequently form a vertical sequence. In contrast, a period of time in which an object is moving is represented in the chart by a sequence of dots forming a slanted line. For example: at 2 seconds after noon, Pete's bullet, just having been fired, is in front of the left post-office window. One tenth of a second later, it has moved 1/5 of the distance to the right saloon window; 5/10 of a second later, it has reached this window; and so forth. The dots representing the position of this bullet at successive intervals of 1/10 second consequently appear not along a vertical but along a slanted line.

The chart in Figure 24 consequently shows that between noon and 1 second after noon Dick is walking to the right and Pete to the left (they are walking toward each other). Then both stop moving; Dick fires, at 2 seconds after noon Pete fires; Dick's bullet and Pete's bullet pass each other at 2.5 seconds after noon, etc. The reader should examine the chart on page 20 carefully, and be sure that he sees how all the facts just stated are represented there. Without a clear understanding of how such charts are to be read, it is useless to go on.

The position of an object at any given instant is, naturally enough, represented by a single point in our chart. Conversely, a point in our chart represents both the time *and* the position in (one-dimensional) space of a single event. The position in (one-dimensional) space of the event which the point represents governs how far to the right or left in the chart the point appears; the time of the event governs how far up or down in the chart the point appears. Since many successive instants are shown all at once in the chart, the successive positions of an object are represented by a succession of points.

The succession of dots showing the position of a given object at successive instants traces out a path in our chart—conveniently, though purely figuratively, called the "path through time" of the object. A stationary object as well as a moving object has such a "path"; the stationary object a vertical "path," the moving object a slanted "path." The reader should avoid confusing this sort of figurative "path" in a chart of events in space and time with the ordinary notion of path through space. In what follows we shall often speak of the "path" of an object in this sense, i.e., in the sense of the path in a diagram representing the successive positions of the object at a large number of instants. The reader, forewarned, should avoid confusion as to what is meant.

Notice that we are putting all events which we determine to have occurred at the same time on the same horizontal line, and putting all events which we determine to have occurred at the same place on the same vertical line. Remembering these two facts, we realize that the horizontal and vertical lines are just guide lines which we might as well leave out, except for a few which we may want to keep for the sake of orientation. Moreover, instead of drawing a more and more closely spaced series of dots showing the successive position of, say, Pete's bullet, at successive intervals of time, first at intervals of 1/10 second, then at intervals of 1/100 second, etc., it is simpler and more informative to let these more and more closely spaced dots "run into each other," and to draw a continuous line which will then show the position of Pete's bullet at every instant of time. With these new stipulations, our picture appears as

Fig. 25

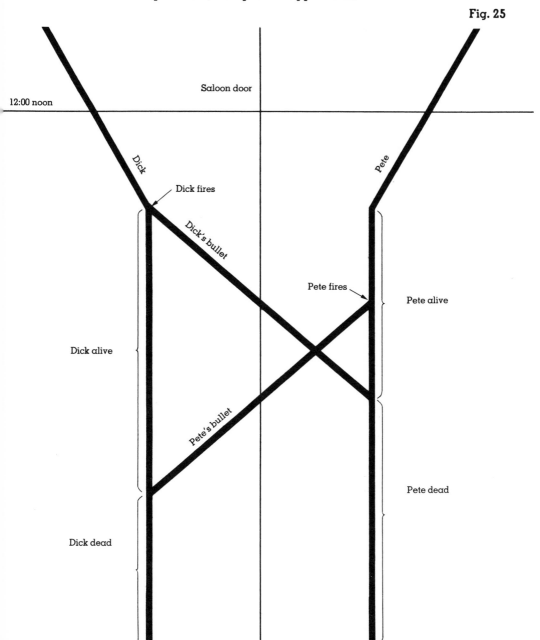

Saloon door

12:00 noon

Dick

Dick fires

Dick's bullet

Pete

Pete fires

Pete alive

Dick alive

Pete's bullet

Pete dead

Dick dead

This is the way we shall draw our simplified "film-strips" from now on. It is a good idea to look back over pages 10, 11, 17, 19, 20, and 23 to see how these simplified drawings have been evolved, and to be sure of what they mean.

The explanation of the way in which Figure 24 is to be read is particularly relevant. Figure 25 is to be read in exactly the same way, except that, whereas Figure 24 shows the situation only at intervals of 1/10 second, the present diagram shows the situation also at, say, 2.15 seconds or 2.1583 seconds after noon. The reader, before going on, should be sure that he can read charts like Figure 25, and that he understands exactly why Figures 3–8, 21, 23, 24, and 25 all give schematic representations of the same sequence of events in various systems of drawing.

It is to be noted that our diagrams show all at once the situation at a whole range of different instants in time; and that the "path" of an object, as it appears in such a diagram, shows the position of the object at a range of different instants, as is explained on page 22.

WARNING:

The continual occurrence of two-dimensional *charts* should not lead the reader to forget that our objects are moving, like trains along a railroad track, in *only one dimension*. The charts are two-dimensional only because an extra dimension is needed to show the *time* at which events occur in our *one-dimensional* space. Dick's bullet, in our chart, is moving in one dimension, from left to right, and *not* moving along a slanted line in two dimensions. In the same way, a film actress who walks from left to right is represented on the actual celluloid film by a series of images which lie along a slant in the celluloid strip; but the actress is not walking at an angle down the celluloid but merely walking, at a certain rate, from left to right on the stage.

We can, if we like, start our "filmstrip" before 12:00 noon to show some of the earlier action.

[The horse happened to pass by just before Pete and Dick made their appearance on the scene. We did not see this in our previous charts simply because none of the earlier action was shown.]

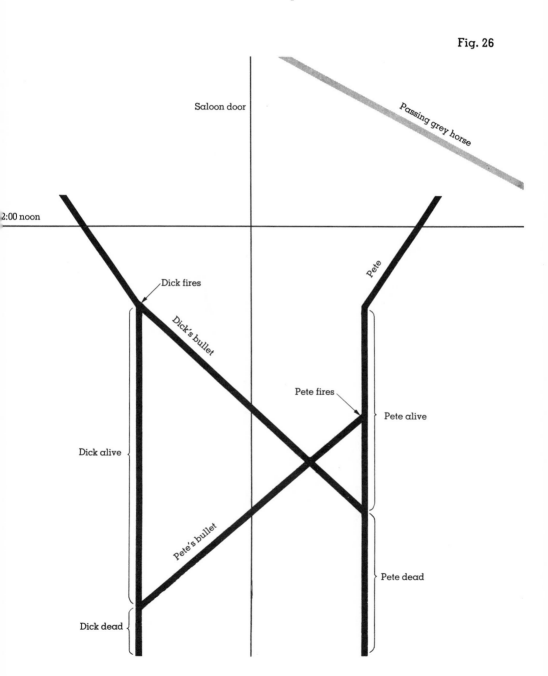

Fig. 26

The picture on the last page is correct if Dick really did fire first. If Pete fired first, it should look like this:

Fig. 27

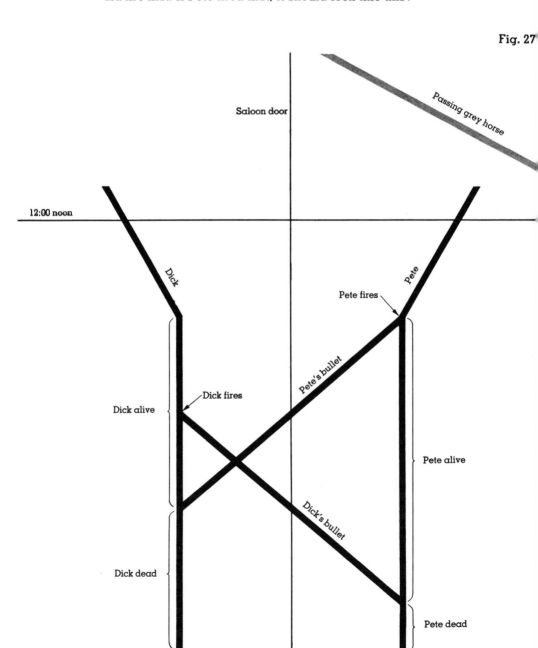

If they both fired at the same time, the picture should look like this:

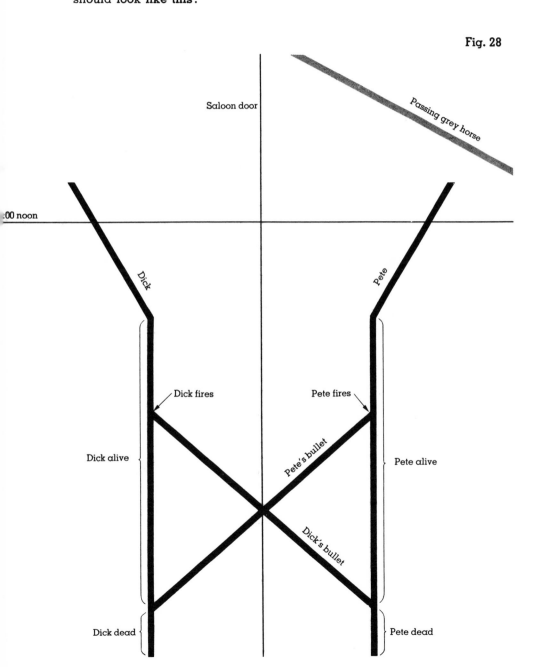

Fig. 28

The three charts in Figures 26, 27, and 28 show the same events pictured as cartoons on pages 10–11, 12–13, and 14–15, respectively. The reader should be sure that he understands the way in which the information in the cartoons is represented in the charts. In our subsequent discussion we shall have continual recourse to charts like those in Figures 26, 27, and 28; a clear understanding of the way in which these diagrams are to be read is vital.

In all the last series of pictures we have followed the simplest and most natural method of drawing, putting all events which we determine to have happened at the same place on the same vertical line, and all events which we determine to have happened at the same time on the same horizontal line.

This, however, *is nothing but a conventional agreement;* we might as well have agreed to put all events which we determine to have happened at the same place on lines sloping up to the right at 40 degrees to the vertical, and to put all events which we determine to have happened at the same time on lines sloping down to the right at 35 degrees to the horizontal. The events pictured in Figure 28 would then appear as follows:

Fig. 29

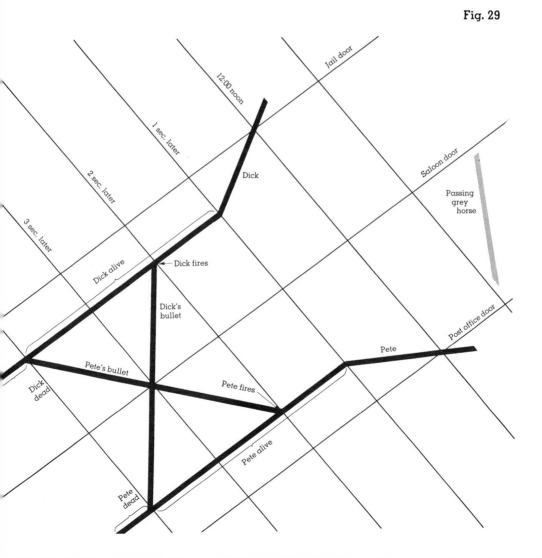

Diagrams like Figure 29, drawn according to a slanted-lines convention instead of a horizontal-vertical-lines convention like that used in the diagram in the preceding figure, are equivalent to diagrams of the horizontal-and-vertical-lines system, even though not as convenient. They have simply to be read a little differently. The difference is as insignificant, as much a matter of convention, as the difference between the English system of writing from left to right, the Chinese system of writing from top to bottom, and the Hebrew system of writing from right to left. Thus, if we look along the slanted line in Figure 29 which represents all the events occurring at 3 seconds after 12, we find that Dick is standing in front of the right jail window, Dick's and Pete's bullets are passing each other in front of the saloon door, and that Pete is in front of the left post-office window. Exactly the same information can be read off Figure 28, except that now it must be read off the horizontal line which, in Figure 28, represents all the events occurring at 3 seconds after 12.

If there were any reason to do so, we could even put all events which we determine to have happened at the same place on a saw-tooth line, and all events which we determine to have happened at the same time on a wavy line. The events pictured on pages 27 and 29 would then appear as follows:

Fig. 30

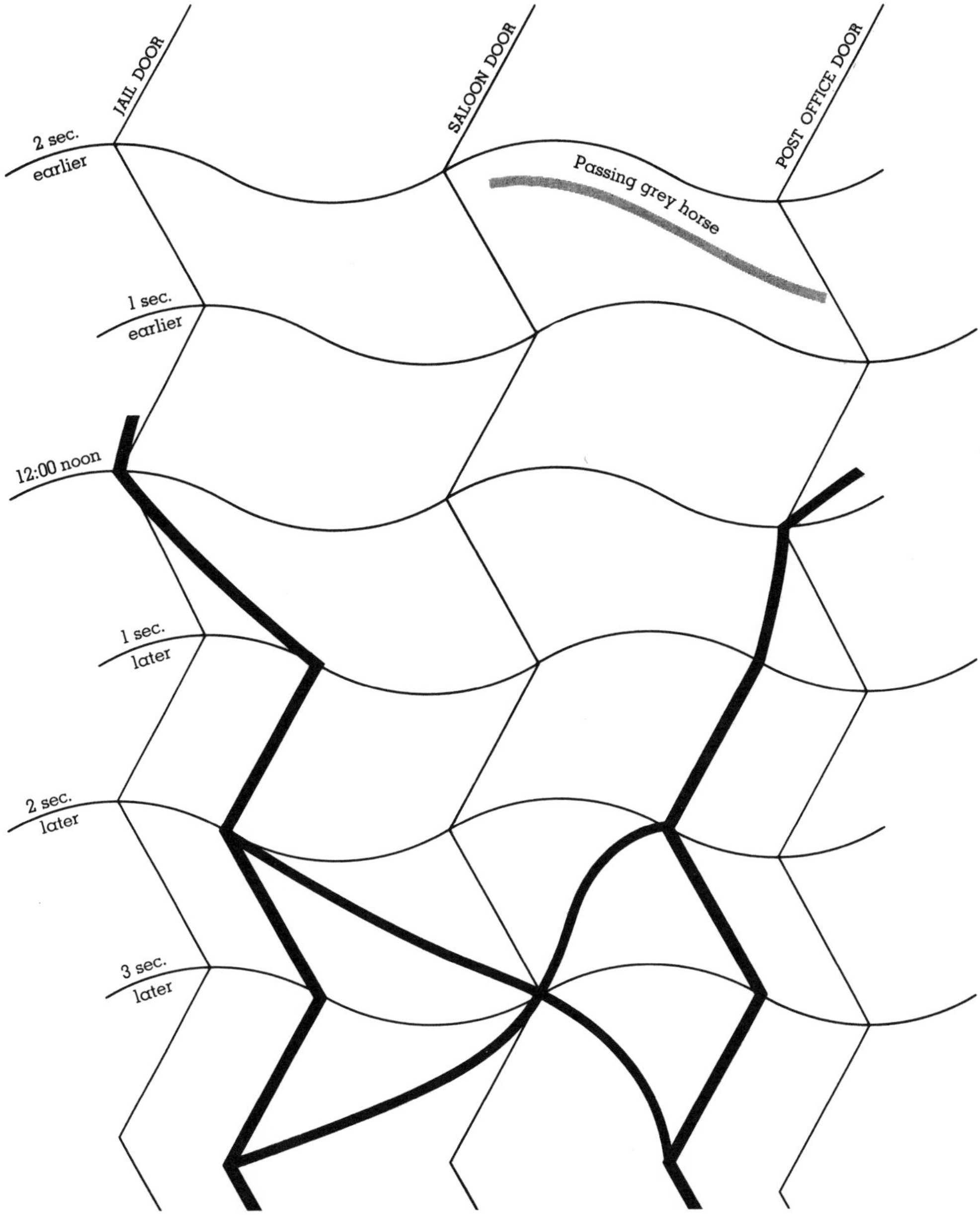

This diagram also shows exactly the same events as Figures 28 and 29. Thus, if we look along the wavy line in the diagram representing all the events occurring at 3 seconds after 12, we find that Dick is standing in front of the right jail window, that Dick's and Pete's bullets are passing each other in front of the saloon door, and that Pete is in front of the left post-office window. This is, of course, the same information represented in the previous diagrams in a slightly different way.

Of course, even though we can draw the picture this way, there is good reason not to: it is confusing. More particularly, our reason for not wanting to draw the picture this way is the same as the reason, given earlier, which determines our basic quantitative assignments of position in space and moment in time to events: we wish our quantitative notions of space and our quantitative notions of time to fit together in such a way that many physical happenings have a simple description. When this is achieved, we know that we have successfully chosen our notions of time and space in a way appropriate for the understanding of the physical world.

What is a typical physical happening with a simple description? One of the very simplest is the motion of a particle on which no forces act, according to the well-known law (sometimes called Newton's first law of motion):

A particle moving without outside influence continues to move indefinitely with the same speed and direction that it has originally.

Thus, if an isolated particle moves 1/2 foot to the right in the 1st second of observation, it moves 1/2 foot to the right in the next second, in the 3rd second, the 4th, etc. The picture of its motion is simply:

Fig. 31

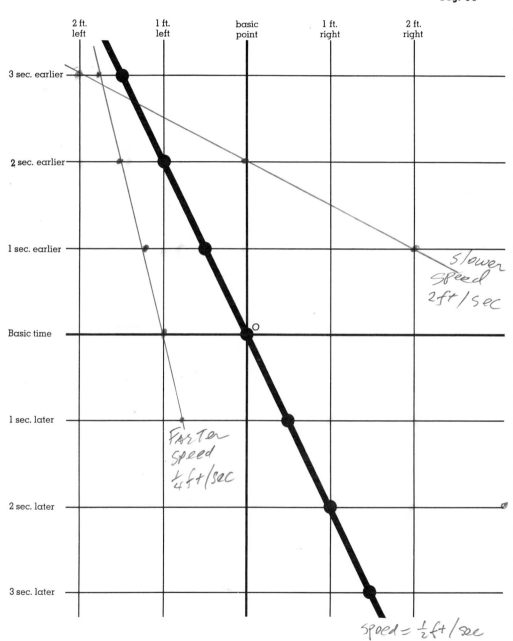

2 ft. left   1 ft. left   basic point   1 ft. right   2 ft. right

3 sec. earlier

2 sec. earlier

1 sec. earlier

Basic time

1 sec. later

2 sec. later

3 sec. later

Slower speed 2 ft/sec

Faster speed 1/4 ft/sec

speed = 1/2 ft/sec

It should be noted that in Figure 31 the position of the particle at successive instants is represented by points which lie along a straight line. This is because the motion is uniform, the speed unvarying. Each second the particle moves 1/2 foot; in our chart, looking down one unit means that the point is to be found 1/2 unit to the right of its former position. If the motion of the particle were not uniform but accelerated, the position of the particles at successive instants would be represented by points lying along some curve. *Uniform motion* is represented in our charts by a slanted straight line; *absence of motion* by a vertical straight line; *varying motion* by a curve.

These facts are extremely convenient: since we shall generally have to consider only uniform motion, we shall generally have to deal only with straight lines, and consequently can use the very simplest geometry.

[*Reminder:* It should again be emphasized that the particle whose successive positions are portrayed in the preceding diagram moves, like all our particles, in one dimension; in the present case, from left to right. The speed with which it moves governs the slope of the line in the chart representing its position at successive instants: the greater the speed, the larger the slope. It must be remembered that the particle is moving *left or right in one dimension* with a greater or smaller speed, and *not* moving down a slanted line in the two-dimensional chart. The additional, vertical, dimension in the chart simply reflects the fact that we show the successive positions of the particle at many instants of time all at once.]

Figure 31 was drawn using the convention that all events determined to have occurred at the same place are put on the same vertical line, and all events determined to have occurred at the same time are put on the same horizontal line. The appropriateness of this system appears in the fact that the uniform motion of a particle moving without outside influence appears as a simple straight line. If instead we used the inappropriate convention of Figure 30, the motion of the particle would have to be represented by a complicated curve.

The appropriateness of our basic notions of time and space appears in the same way. Our particle, moving without outside influence, travels the same number of feet each second. If we used an inappropriate system of time and space assignments, the particle would seem to go a greater distance in some seconds and a smaller distance in others. This complicated description corresponds exactly to the impressions of someone whose watch runs badly, or who measures with a ruler whose marks are not evenly spaced. For such a person Newton's very simple law of motion given on page 32 has to be replaced by an extremely complicated specification of the fact that in some particular seconds (as determined by a bad watch or other inappropriate system of time assignments) isolated particles travel a greater distance, and in some particular seconds a lesser distance. We know that our system of assigning positions in space and moments in time to events is *right,* and the system of the man with the bad watch *wrong,* by the fact that we find a simple law for the motion of an isolated particle, while he can find only a complicated law. If he is reasonable, we can even use this fact to convince him of his error.

The law of motion stated on page 32 is related to an even more far-reaching law, whose consequences we have all experienced many times, though perhaps without realizing what is involved. When we travel along in an automobile, ship, airplane, or railroad car, then, to the extent that the motion of our conveyance is steady (so that we feel no "bumps"), we do not feel any effects of the motion at all. All the laws of physics to which we become accustomed while "standing still" serve us just as well while we are in uniform motion: aboard a uniformly moving conveyance one can walk, stand, sit, throw and catch objects, etc. just as if one were not moving at all. Of course, things are quite different when the conveyance is not moving uniformly but is changing its speed, that is, either starting or stopping. Then we feel very definite effects and must brace ourselves to keep from falling forward or backward, etc. In the event of a very sudden halt the effects of a change of motion can become so great as to be catastrophic. But, as long as the motion of a conveyance continues uniformly there are no effects at all. This familiar fact is called the *GENERAL LAW OF UNIFORM MOTION* and can be expressed as follows:

An observer in uniform motion finds all the same laws of physics to hold as does a stationary observer.

Because of this law it doesn't make much sense to speak of one observer as "really" stationary and of another as "really" moving uniformly. Two such observers have exactly corresponding experiences; they both find the same laws of physics to be operating. There is no way in which one would be made conscious of motion any more than the other, no way in which either could prove or be convinced either that he was really stationary or that he was really moving. Each uniformly moving observer is consequently free to consider, without making any error, though really only as a matter of convention, that he personally is the one who is stationary while the other observers move past him.

Of course, the case of an observer who is not moving uniformly but whose motion is changing is quite different. A man who has been through an automobile crash has for a short while had experiences quite unknown to either a stationary or a uniformly moving observer. He can readily be convinced that his state of motion has undergone a change by the fact that the windshield of his automobile seemed suddenly to hit him in the face. The same applies to a pedestrian who is hit by a car: he knows also that his state of motion has not been uniform but has changed. While mere uniform motion, according to the General Law of Uniform Motion, makes no difference, bumps in the motion certainly do.

All these facts have been known for hundreds of years. They were stated clearly by Galileo, who wrote in 1630 in his *Dialogue on the Two Great World Systems,* in first quoting and then refuting those who would prove the earth not to rotate but to be stationary:

> ... **All allege that heavy bodies moving downwards move by a straight line perpendicular to the surface of the earth, an argument which is held to prove undeniably that the earth is immovable. For otherwise a tower from the top of which a stone is let fall, being carried along by the rotation of the earth in the time that the stone spends in falling, would be transported many hundred yards eastwards, and so far distant from the tower's foot would the stone come to the ground. This effect they back with another experiment: letting a bullet of lead fall from the round top of a ship at anchor....If the same bullet be let fall from the same place when the ship is under sail, it will light as far from the former place as the ship has sailed in the time of the lead's descent. ...They add a third and very evident experiment, namely, that shooting a ball point-blank out of a cannon towards the east, and afterwards another with the same charge and at the same elevation towards the west, the range towards the west should be very much greater than the other towards the east ... [but], mark you, whosoever shall perform it shall find the event succeed quite contrary to what has been written of it. That is, he shall see the stone fall at all times in the same place of the ship, whether it stand still or move with any velocity whatsoever. So that, the same holding true in the earth as in the ship, one cannot, from the stone's falling perpendicularly at the foot of the tower, conclude anything touching the motion or the rest of the earth.**

One of the simplest proofs that mere uniform motion cannot be detected through any physical effect comes from our own motion with the earth around the sun. We travel this distance of 586,000,000 miles in 365 days, therefore traveling approximately 1.6 million miles a day, hence 66,000 miles an hour, or 19 miles a second.

Traveling as we must, with this speed of 19 miles a second, we detect—nothing: no effects at all. But if any physical effects of a uniform velocity existed, the rather considerable velocity of 19 miles a second, i.e., 66,000 miles an hour, should be sufficient to make these effects noticeable. Since we detect nothing, we may claim to have powerful evidence for the truth of the General Law of Uniform Motion.

An observer in uniform motion finds all the same laws of physics to hold as does a stationary observer.

Thus, working from his own physical experience, a uniformly moving observer arrives at quantitative notions of time and space as much justified as those of a "stationary" observer: indeed, there is no objective basis on which to distinguish one observer as moving and the other as stationary. Each uniformly moving observer is free to consider, without making any error, that he personally is stationary, while others move past him.

Now, how are the quantitative notions of space and time of one uniformly moving observer related to those of another uniformly moving observer? Which events does each consider to occur at the same place, and which at the same time?

Here we come, for the first time, to the heart of our subject. Having established the system of diagrams which are to be our language, and having the principle of uniform motion as our essential key, we are in a position to proceed.

How are the quantitative notions of space and time of one rational and scientifically accurate uniformly moving observer related to those of another rational and scientifically accurate uniformly moving observer? Which events does each consider to occur at the same place, and which at the same time?

To answer these questions, we must remember that each observer's quantitative notions of time and space come from his physical experience. Thus, to answer the questions which have just been asked, we must consider the details of the physical experience of each of the two observers.

The reader should bear in mind the fact that we take all observers to have "correct" notions of time and space in the sense of pp. 34–35. Moreover, each observer, while considering himself to stand still and the others to move, realizes that the others consider themselves to stand still and him to move.

It should be noted before anything else that the two observers, and indeed all observers, will agree qualitatively about what occurs. They will agree, for instance, that certain particles existed and moved for a while, and then collided with each other, some disappearing and others coming into existence; that certain explosions and flashes of light occurred; etc. What they may or may not disagree about is the location in space and the time of these events, and about which occurred first and which occurred afterward. *Since Mr. A and Mr. B are moving relative to each other, they naturally come upon the events which constitute their common physical experience in a somewhat different order. Thus it is entirely plausible that the notions of time and space which Mr. A elaborates from his physical experience, and which, as a consequence, objectively govern Mr. A's experience, could be somewhat different from the notions of time and space which Mr. B elaborates from his physical experience, and which, as a consequence, objectively govern Mr. B's experience. The actual extent, if any, of their quantitative disagreement is what we must investigate.*

Let us take the point of view of one of the two uniformly moving observers, call him Mr. A, and let us consider him (as he considers himself) to be stationary. He represents his own notions of space and time in a chart, putting on a vertical line all events deemed to occur at the same place, and on a horizontal line all events deemed to occur at the same time. A uniformly moving observer, call him Mr. B, passes by this fixed observer. As we see from page 33, the successive positions of the uniformly moving observer are represented by a straight line. Thus, in the representation employed by the stationary observer, the whole situation looks as follows:

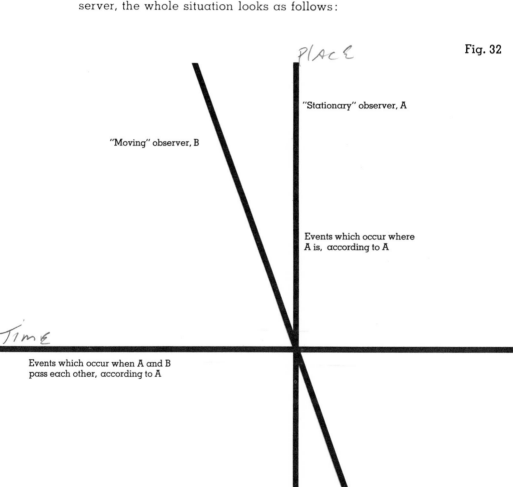

PLACE

Fig. 32

"Stationary" observer, A

"Moving" observer, B

Events which occur where A is, according to A

TIME

Events which occur when A and B pass each other, according to A

Mr. A asks himself: Which sets of events does Mr. B consider all to occur at the same place, and which sets of events does Mr. B consider all to occur at the same time? How do Mr. B's notions of time and space appear when represented in my chart along with my own notions of time and space?

Common sense steps in with the immediate answer: Since time is the same for everybody, and since Mr. A and Mr. B are both competent observers making no mistakes, they must agree on the question of whether a number of events all take place at the same time. Thus, common sense would say: Since Mr. A has represented all events taking place at the same time by the points on a horizontal line in his chart, Mr. B must also agree that it is the points on this horizontal line which represent events taking place at the same time. Thus, for common sense, the question is no sooner asked than answered.

We shall see in the course of our discussion that the answer given by common sense is wrong. As has been said, it is certainly *plausible* that it should be wrong, for we have seen that Mr. A and Mr. B both elaborate their notions of time and space from their physical experience. Since Mr. B and Mr. A are moving relative to one another, they come upon the events which make up their common experience in somewhat different orders. Do their notions of time and space agree anyhow, as common sense insists? Let us take a crucial step of intellectual self-liberation by admitting frankly: At this stage of the discussion we do not know. We do not know but wish to find out. Let us investigate the question as carefully and precisely as we can. It may be that at the end of our investigation we shall find that common sense was right after all. If so, fine. We shall then understand why it was right. It may be, on the other hand, that we shall find that common sense is wrong.

Mr. A then asks: What are Mr. B's notions of time and space? Mr. A's own notions of the time and place **at which events occur are represented in Mr. A's chart. Mr. A wants to know, which of these events does Mr. B consider to occur at a given time? How do** Mr. B's notions of time and space appear in Mr. A's chart? If Mr. A could succeed in representing Mr. B's notions of time and space in his own chart, it would be easy to tell at a glance whether or not Mr. B's notions agreed with Mr. A's. But how to begin? We have agreed not to accept uncritically the answer given by common sense. Mr. A prepares a chart representing his own notions of space and time, putting on a vertical line all those events deemed by himself to occur at the same place and on a horizontal line all those events deemed by himself to occur at the same <u>time</u>. But, as far as Mr. A knows, Mr. B might consider a rather odd set of events to occur at the same time, say, all the points marked with an asterisk in Mr. A's chart as follows:

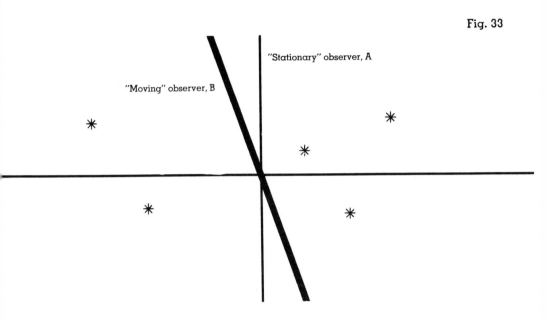

Fig. 33

"Stationary" observer, A

"Moving" observer, B

Or all the points lying along a curve in Mr. A's chart as follows:

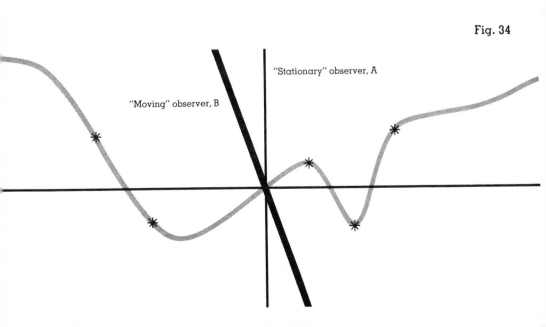

Fig. 34

"Stationary" observer, A

"Moving" observer, B

How can firm footing be found in the apparent infinitude of possibilities which might occur once we suspend faith in the easy answer of common sense? Fortunately, the range of possibilities may be cut down drastically by the following arguments. According to the General Law of Uniform Motion, both observers find the same laws of physics to hold. Among these laws is the law governing the motion of a particle not subject to outside influence:

A particle moving without outside influence continues indefinitely to move uniformly with its original speed.

Now, Mr. A and Mr. B will agree on whether or not a particle is isolated, since if anything is near the particle to exert a force on it, both Mr. A and Mr. B would be aware of this. Since Mr. A knows the particle to be isolated, he knows that Mr. B also considers it to be isolated. As we saw on page 33, the motion of such an isolated, and hence uniformly moving, particle, is represented by a straight line. *Thus, Mr. A realizes that, whatever Mr. B's notions of space and time are, both he and Mr. B must agree on what sets of events are properly represented in a chart of events as lying together on a straight line and what sets of events are not.* Moreover, Mr. A knows that, in Mr. B's chart of events, Mr. B will represent all events which Mr. B deems to take place at the same point by vertical straight lines and all points which Mr. B deems to take place at the same time by horizontal straight lines. Since we have just seen that Mr. A knows that he and Mr. B agree on what sets of events are to be represented by straight lines (even though they need not necessarily agree on whether the straight lines are to be horizontal, vertical, or slanted), Mr. A must also know that the events which Mr. B considers to occur at a given place lie along certain straight lines *IN MR. A's OWN CHART*, and the events which Mr. B considers to occur at a given time lie along certain other straight lines in Mr. A's chart.

In Mr. A's chart Mr. B's notions of time and space would consequently have to appear something like this:

Fig. 35

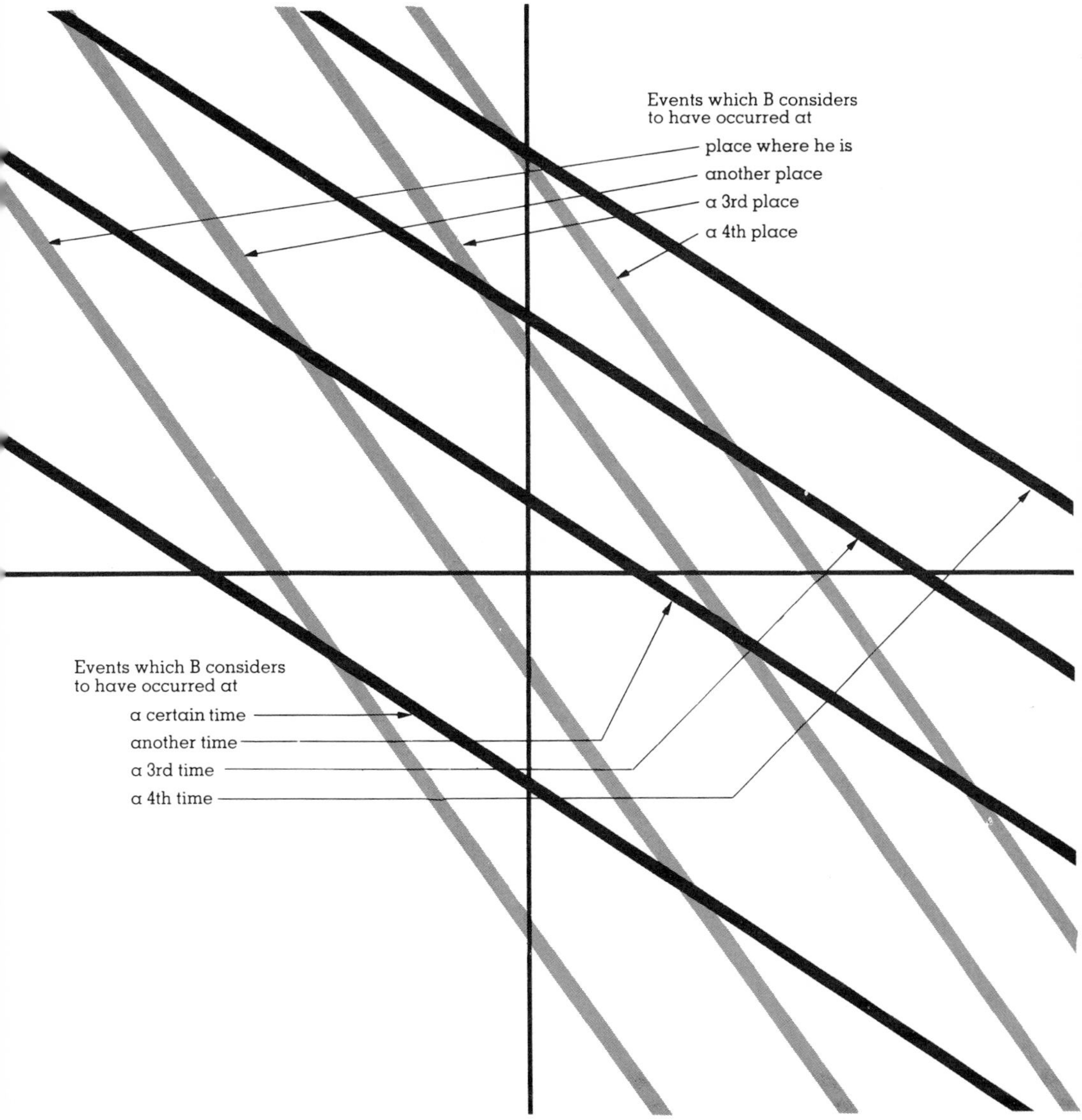

Events which B considers to have occurred at

place where he is
another place
a 3rd place
a 4th place

Events which B considers to have occurred at

a certain time
another time
a 3rd time
a 4th time

Let us review verbally the information given pictorially in Figure 35. The events which Mr. B considers to occur at a given place must lie along a straight line, and not any more complicated curve, in Mr. A's chart. Likewise, the events which Mr. B considers to occur at a given time must lie along a straight line in Mr. A's chart. If a straight line L in Mr. A's chart represents all the events which Mr. B considers to occur at a given place, and a straight line M in Mr. A's chart represents all the events which Mr. B considers to occur at some other place, then these two straight lines must be parallel, for in Mr. B's chart these sets of events are both represented by vertical lines, and hence have no point in common; their intersection, if any, would be a *single* event that Mr. B considered to occur both at a place and at a different place. Thus, the family of all lines in Mr. A's chart representing sets of events which Mr. B considers to occur at various fixed places in space must be a family of parallel lines. Likewise, the family of all lines in Mr. A's chart representing sets of events which Mr. B considers to occur at various fixed moments in time must be a family of parallel lines.

Mr. A also realizes that Mr. B considers himself to be stationary. Hence, Mr. B must consider the events which occur at the points where Mr. B is at successive instants to have occurred at the same place. These events are exactly the events which happen along the dark gray straight line in Figure 32; this dark gray line simply shows where Mr. B has been.

The path of Mr. B is a line, like the dark gray lines in the diagram on page 45, which represents all those events which Mr. B considers to occur at a single definite place (to wit, the place where he is). Consequently, all these dark gray lines must be *parallel to Mr. B's path*, so that the picture on page 45 can be made a little more specific; it must actually look like this:

Fig. 36

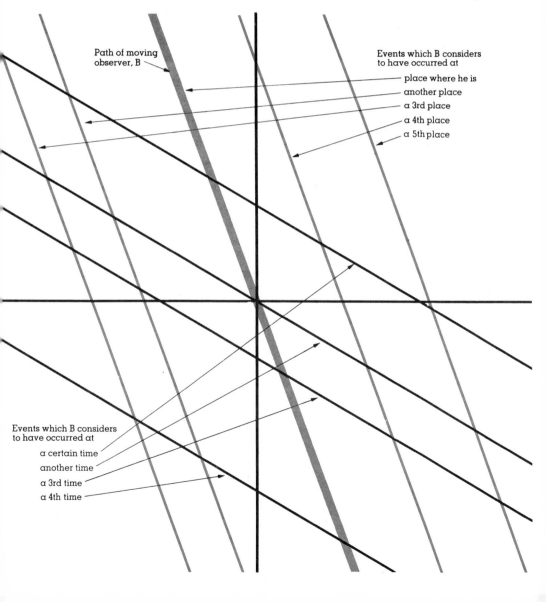

Path of moving observer, B

Events which B considers to have occurred at

place where he is
another place
a 3rd place
a 4th place
a 5th place

Events which B considers to have occurred at

a certain time
another time
a 3rd time
a 4th time

To summarize: The family of all lines in Mr. A's chart which represent sets of events which Mr. B considers to occur at various fixed points in space is a family of lines parallel to the line representing the successive positions of Mr. B. The family of lines in Mr. A's chart which represent sets of events which Mr. B considers to occur at various fixed moments in time is another family of parallel lines.

These assertions give basic information on the relationship between Mr. A's notions of time and space and Mr. B's notions of time and space. To pin down this relationship more precisely, we must answer such additional questions as: What angle do the lines in Mr. A's chart which represent sets of events which Mr. B considers to occur at various fixed moments in time make with the horizontal? What is the proper spacing between these lines? And so forth.

Our aim is to elucidate these questions.

Mr. A can say something about the spacing of the straight lines which represent Mr. B's notions of "fixed time" and "fixed place" by using the following reasoning. Suppose that two particles, both not subject to any outside influence, move along from right to left with the same unchanging velocity, so that they are always the same distance apart, as Mr. A reckons distance. In Mr. A's chart the motion of these two particles would appear as follows:

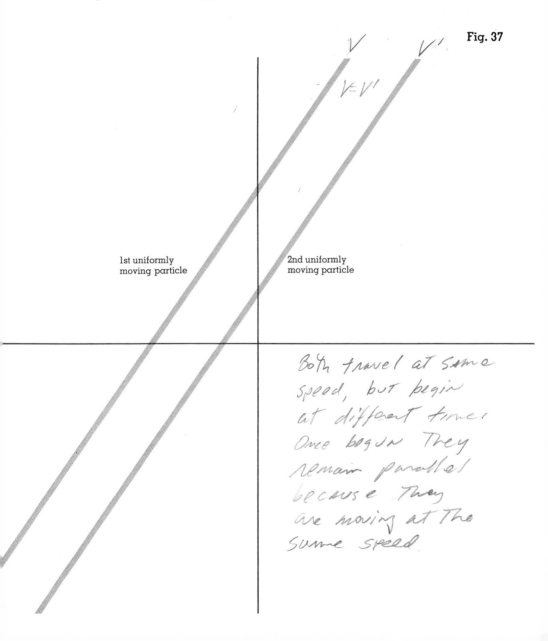

Fig. 37

1st uniformly moving particle

2nd uniformly moving particle

Both travel at same speed, but begin at different times. Once begun They remain parallel because They are moving at The same speed.

Since Mr. A considers the two particles to be moving with the same speed, the successive positions of the particles are represented in Mr. A's chart by two lines of equal slope, hence by parallel lines. Another way of seeing that these two lines must be parallel is as follows: If the lines were not parallel, they would intersect. The point of intersection would represent a time and place at which the two particles come together. But, since they are traveling with the same speed (according to Mr. A), they stay the same distance apart, and hence certainly never come together.

Now, Mr. A knows that Mr. B agrees with him as to which motions are to be represented by straight lines, i.e., agrees as to which motions are uniform. Hence, he knows that Mr. B also considers each of these two particles to be moving with a fixed velocity; one with velocity v, the other with velocity v'. We shall now establish that Mr. B must consider these two velocities to be the same, i.e., that $v = v'$. This will be done by the method of indirect proof, i.e., by showing that the supposition that v and v' are different leads to a consequence which is demonstrably false. Suppose, then, that the two velocities v and v' are different.

We note that, since we are considering motion in one dimension only, v and v' are the velocities with which Mr. B takes the particles to move in their one-dimensional space (along their railroad track, if you will). We shall suppose, for the sake of definiteness, that v and v' are the velocities with which Mr. B takes the particles to move leftward and that v is the greater of v and v'. There are then two possible cases.

Case 1. The particle with the larger velocity v (as judged by Mr. B) is to the right of the particle with the smaller velocity v'. But then the former particle is overtaking the latter, and will catch up with it after the lapse of a certain period of time. Thus, there is a time and place (in the future, according to Mr. B) at which the two particles come together.

Case 2. The particle with the larger velocity **v** (as judged by Mr. B) is to the left of the particle with the smaller velocity **v'**. But then the former particle is getting farther and farther ahead of the latter, and hence, since the velocities are constant, the two must have been at the same place at some time (in the past, as judged by Mr. B).

At any rate, then, if Mr. B deems the two velocities v and v' to be different, he must believe that there is some time at which the two particles are at the same place. We know, however, that Mr. A and Mr. B must agree qualitatively as to what events happen, even though they may disagree about the order in space and time of these events (see page 39). Hence, if Mr. B thinks that the two particles whose motion is shown on page 49 are ever at the *same* place at the *same* time, i.e., that it ever happens that these particles pass each other, Mr. A must agree that these particles pass each other at some place and time, i.e., that at some particular instant in time they are at the same place. However, it is perfectly clear that since Mr. A considers these two particles to have the same velocity, he can never expect them to pass each other; the same is also clear from Figure 37—Mr. A does not consider that the two particles are ever at the same place at the same time (since the two light-gray lines never intersect). Thus, the assumption that v and v' are different has a demonstrably false consequence. Hence, Mr. A can conclude that Mr. B must agree with him that these two particles have the same velocity. Using this basic fact, Mr. A may now proceed to discover the spacing between the lines representing successive moments of time as time is regarded by Mr. B.

The discussion above has shown that Mr. B must agree with Mr. A that the two particles whose motion is depicted in Figure 37 have exactly the same velocity. Suppose that we draw, in Mr. A's chart,

both Mr. B's path and the line representing all those
events which Mr. B considers to occur at the instant
that he passes the first of the two particles. The resulting picture would look like this:

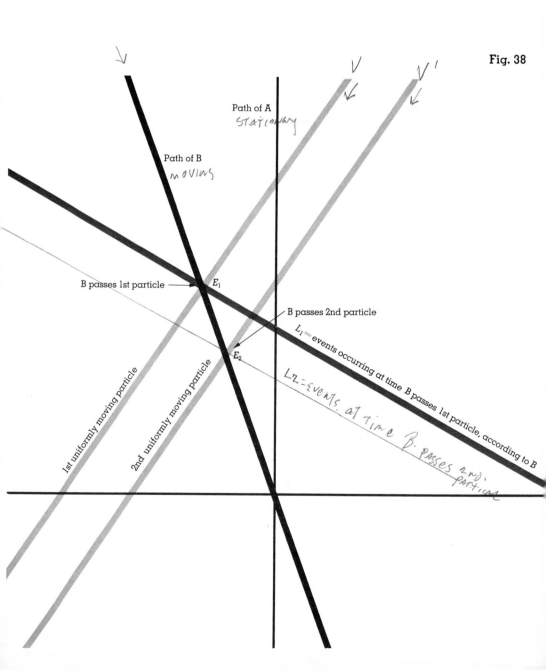

Fig. 38

Let the common velocity which Mr. B ascribes to the two moving particles be called v and the distance between the two particles be called d; then $t=d/v$ is the time it takes the particles to traverse the distance d. Hence, it will be exactly t seconds after the first particle passes Mr. B (as Mr. B reckons time) that the second particle passes him. Thus Mr. B must consider the event E_2 in the above chart to occur exactly t seconds after the event marked E_1. Consequently, a line L_2 drawn through E_2 parallel to the line L_1 will represent all those events which, according to Mr. B, occur t seconds after Mr. B passes the first particle. The intersection of this line with the path of the first particle represents the point which this particle

$d = $ distance between particles
$v = $ velocity of particles
$t = $ time

reaches t seconds after Mr. B passes it, as Mr. B reckons time. Hence, this intersection must represent an event d feet to the left of Mr. B, as Mr. B reckons distance. This is shown on the following diagram.

Fig. 39

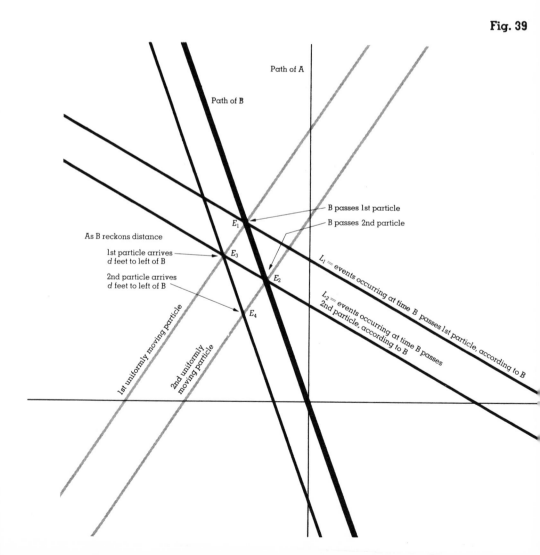

The line parallel to the path of Mr. B through the point E_3 shows all those events which, according to Mr. B, occur d feet to the left of Mr. B. The intersection E_4 of this line with the path of the second particle shows the arrival of the second particle d feet to the left of Mr. B. This event naturally occurs t seconds after the second particle passes Mr. B, as Mr. B reckons time.

We may now repeat our argument over and over again. We have seen that the event E_2 occurs t seconds after the event E_1, as Mr. B reckons time, and that the event E_4 occurs t seconds after E_2. If we draw a line L_3 through E_4, parallel to lines L_1 and L_2, it will represent all those events which, according to Mr. B, occur t seconds after E_2. The intersection E_5 of line L_3 and the path of the first particle will show the arrival of the first particle at a point $2d$ feet to the left of Mr. B at the moment the second particle arrives d feet to the left of Mr. B, as Mr. B reckons time and space. Thus the line through E_5 parallel to Mr. B's path shows all those events which, according to Mr. B, occur $2d$ feet to the left of Mr. B's position. (See Figure 40.)

The intersection E_6 of this line and the path of the second uniformly moving particle show the arrival of the second particle $2d$ feet to the left of Mr. B, and hence show an event that must occur t seconds after E_4. Thus the line L_4 through E_6, parallel to the lines L_1, L_2, and L_3, must represent all those events that, according to Mr. B, occur t seconds after E_4.

Reasoning repeatedly in this way, Mr. A, who knows that Mr. B's notions of space and time are qualitatively as represented on page 47, deduces that quantitatively they must be as represented by the following picture:

Fig. 40

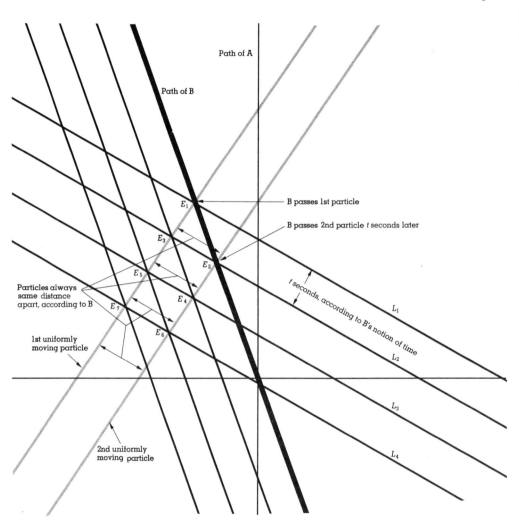

That is, the lines representing all those events which, according to Mr. B, occur at places d feet apart, etc. are evenly spaced in Mr. A's chart; and the lines representing all those events which, according to Mr. B, occur at a given time, t seconds later, twice t seconds later, etc. are also evenly spaced.

The information about Mr. B's notions of time and space recorded by Mr. A in the last picture is so important it is worth our while to draw it once again, without such a clutter of auxiliary lines and captions.

Fig. 41

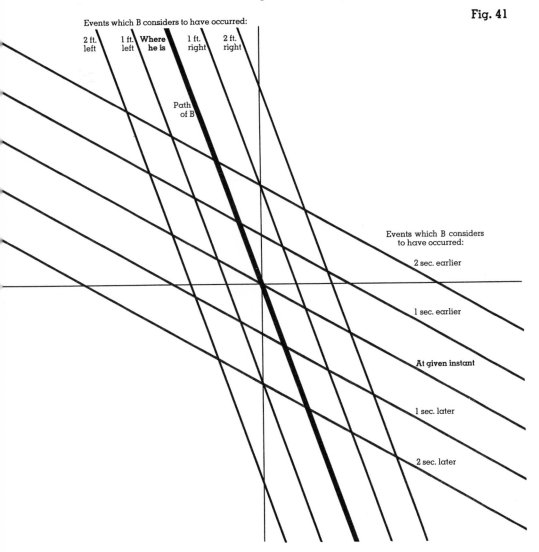

Events which B considers to have occurred:

2 ft. left 1 ft. left Where he is 1 ft. right 2 ft. right

Path of B

Events which B considers to have occurred:

2 sec. earlier

1 sec. earlier

At given instant

1 sec. later

2 sec. later

To restate the information portrayed graphically in Figure 41, for emphasis: The family of parallel lines in Mr. A's chart representing all those events which, according to Mr. B, occur at various places in space separated by intervals of 1 foot, is an evenly spaced family of lines parallel to the line representing Mr. B's successive positions. The family of parallel lines in Mr. A's chart representing all those events which, according to Mr. B, occur at various moments in time separated by intervals of 1 second are also evenly spaced.

Each set of lines is evenly spaced; but for all we know so far, the spacing may be different for each set.

Mr. A would have a complete knowledge of Mr. B's notions of time and space if he could answer just three more questions.

1. At what angle with the vertical are the lines which represent all those events which Mr. B considers to occur at the same time?

2. How large is the gap between the lines which represent all those events which Mr. B considers to occur at a given instant, 1 second later, 2 seconds later, etc.?

3. How large is the gap between the lines which represent all those events which Mr. B considers to occur at a given point, 1 foot over, 2 feet over, etc.?

It is not so easy to give a correct answer to these questions. To do so will be our next task, which will occupy us for the next dozen pages.

Common sense, however, *seems* to give an obvious answer. According to common sense, time is time; the time which governs Mr. A's physical experience, and the time which governs Mr. B's experience, and anybody else's for that matter, is real physical time, which passes in the same way for everybody, objectively, uniformly in the whole universe. Thus, according to common sense, the notion of time which Mr. B arrives at from his physical experience must necessarily be the same as Mr. A's notion of time. Hence the lines about which questions 1 and 2 are asked must be the same horizontal lines which Mr. A considers to represent events occurring at a given instant, 1 second later, 2 seconds later, etc. From the point of view of common sense, it is even foolish to speak in terms of what "Mr. A considers," or "Mr. B considers." These lines represent the sets of events which really, objectively, occur at a given objective instant, 1 second later, 2 seconds later, etc. If either Mr. A or Mr. B considers differently, he is simply mistaken. But, since we assume that both Mr. A and Mr. B have taken correct account of their own physical experience, and of the physical laws which the General Law of Uniform Motion tells us are common to them, neither of them has made any mistake, and hence both must have the same notion of time— the true, objective notion.

So common sense tells us that the picture on page 57 must really look like this:

Fig. 42

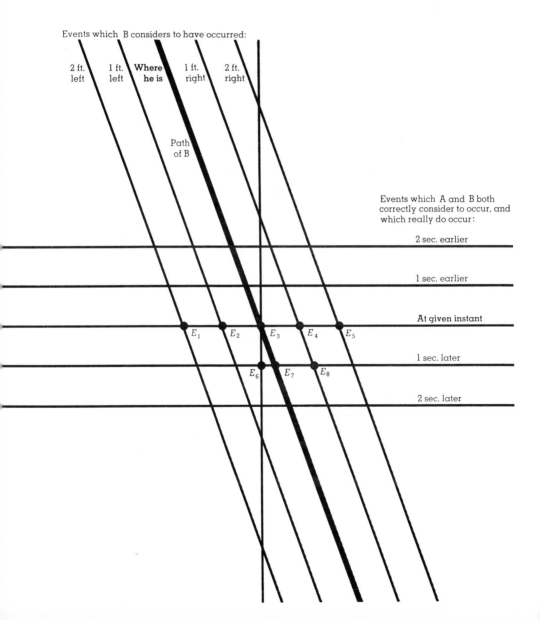

Events which B considers to have occurred:

2 ft. left · 1 ft. left · **Where he is** · 1 ft. right · 2 ft. right

Path of B

Events which A and B both correctly consider to occur, and which really do occur:

2 sec. earlier

1 sec. earlier

At given instant

E_1 · E_2 · E_3 · E_4 · E_5

1 sec. later

E_6 · E_7 · E_8

2 sec. later

In agreement with the statement deduced on page 55, and stated on pages 57–58, the family of parallel lines in Mr. A's chart representing all those events which, according to Mr. B, occur at various moments in time separated by intervals of 1 second, are evenly spaced. But common sense insists that, even more, these lines are horizontal (rather than slanted at any angle, however small) and spaced at intervals which Mr. A would also call intervals of 1 second.

Common sense also tells us that distance is distance—the same, objective, physical distance necessarily governing both Mr. A's and Mr. B's physical experience. Thus, according to common sense, both Mr. A and Mr. B, unless they have made a mistake, must agree that the events E_1, E_2, E_3, E_4, and E_5 in the previous diagram occur at the same time and at places 1 foot apart in space, and that both Mr. A and Mr. B must agree that E_6, E_7, and E_8 occur 1 second later, E_7 and E_8 at places 1 foot apart in space. Of course, Mr. A would say that E_3 occurs at the same place as E_6, while Mr. B would say that E_3 occurs at the same place as E_7. But both Mr. A and Mr. B understand that this difference must naturally arise from the fact that each of them considers himself to be stationary and the other to be moving; i.e., each measures distances by using himself as the "zero point." It is a slight difference in their points of view that is of no particular consequence, since according to the General Law of Uniform Motion no physical experience can exist which would tend to prove one right and the other wrong.

Everything up to page 58 has been deduced rigorously; that which has been written from page 58 to the present page merely represents the customary view. We shall soon see that on these matters the customary view, i.e., common sense, is wrong.

Wrong, even though most people have believed for centuries that time and distance are obvious and absolute facts of the physical world. Wrong, even though this common-sense opinion was upheld by the most eminent of scientists. Newton, for instance, wrote in his *Principia:*

> **Absolute, true, and mathematical time, of itself, and of its own nature, flows equably without respect to anything external—absolute space, in its own nature and without relation to anything external, remains always similar and immovable.**

How much and how little! Yet such was the general view until Einstein realized that this common-sense opinion clashes with our actual physical experience of the velocity of light and is shattered in the collision.

Let us look at the facts!

If we produce a flash of light, the light from this flash will travel out in all directions; and will travel at the same uniform velocity in all directions, arriving as far to the left as to the right in any interval of time. [We note for emphasis that, as was stated on pages 18 and 24 and repeatedly thereafter, we take all motion to transpire in a one-dimensional space, so that all motion is motion either to the left or to the right, and two moving particles are either coming together or moving apart.] Let a uniformly moving observer, Mr. B, pass another uniformly moving observer, Mr. A. We shall look at things from the point of view of Mr. A, and suppose that Mr. A is stationary, while Mr. B moves by from left to right with velocity v, so that 1 second after passing Mr. A, Mr. B is v feet to the right of him. At the instant and place Mr. A and Mr. B pass each other, they arrange that a flash of light shall be produced. The light from this flash travels to the left and to the right with a velocity c. In Mr. A's chart of events all this appears as follows:

Fig. 43

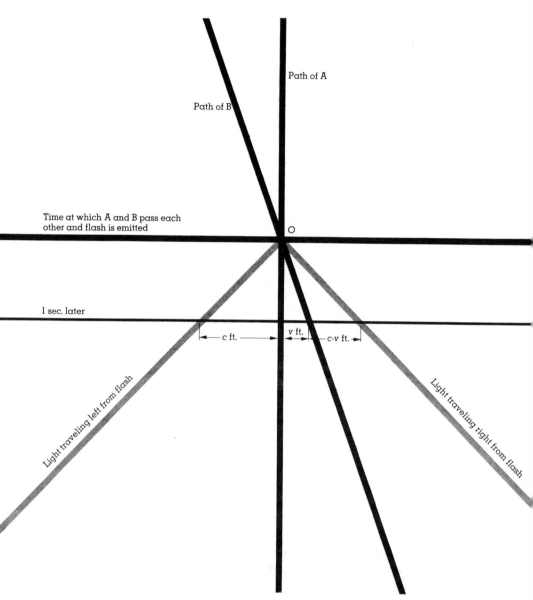

One second after the flash, as Mr. A reckons time, the leftward-traveling light from the flash is c feet to the left of Mr. A, the rightward-traveling flash is c feet to the right of Mr. A, Mr. B is v feet to the right of Mr. A, and consequently the rightward-traveling flash is $c - v$ feet to the right of Mr. B, while the leftward-traveling flash is $c + v$ feet to the left of Mr. B. But then, if common sense is right, if time and space are absolute and objective, so that Mr. B agrees with Mr. A as to which events occur 1 second later and on the distance between these events, then it is plain from Figure 43 that Mr. B will *also* find that 1 second later the part of the light from the flash which is traveling to the left is $c + v$ feet to the left of him, while the part of the light which is traveling to the right is only $c - v$ feet to the right of him. *But this would mean that, as far as Mr. B was concerned, the light from the flash traveled to the left faster than to the right!*

This last conclusion makes perfect common sense. Mr. A sees the light traveling to the right at the same speed that it travels to the left. But Mr. B is traveling to the right. Therefore he is running away from the part of the flash which travels to the left and running after the part of the flash which travels to the right. That part of the flash which he is catching up with must seem to Mr. B to be going more slowly, and that part of the flash that he is going away from must seem to him to be going faster. This is exactly like the experience of driving at 60 miles an hour behind a car going 61 miles; common sense tells us that the car ahead seems to be advancing only 1 mile an hour. A car on the other side of the road, traveling at 60 miles an hour in the opposite direction, passes one at 120 miles an hour.

The common-sense view is that time and distance are absolute realities which appear in the same way to everybody, and that Mr. A and Mr. B must consequently agree that 1 second after the light flash the situation is as is pictured on the previous page, so that if Mr. B runs with a speed v after a flash traveling with speed c he must find the speed of the flash to be $c - v$. This common-sense view is perfectly consistent; all its various consequences support each other. But a view can be consistent without being correct.

Is this common-sense view correct? Suppose it is.

Since we saw that the earth travels in its orbit with a speed of 19 miles a second, a measurement of the velocity of light from a flash should show it traveling in one direction at least 38 miles a second faster than in the opposite direction. *HERE IS A STATEMENT WHICH CAN BE CHECKED EXPERIMENTALLY.*

It was checked experimentally for the first time by Michelson and Morley in the second week of July, 1887. They carefully measured the speed with which light travels in various different directions, and found that *IT TRAVELS WITH THE SAME SPEED IN EVERY DIRECTION.* (They had to be careful, since light travels approximately 186,000 miles per second, and the difference of 40 miles per second which they were looking for is only a twentieth of 1 per cent of this. But they were able to measure to this accuracy.) Thus they discovered *A FACT OF PHYSICAL EXPERIENCE INCOMPATIBLE WITH THE PICTURES ON PAGES 59 AND 62; I.E., INCOMPATIBLE WITH THE COMMON-SENSE NOTION THAT TIME AND DISTANCE ARE ABSOLUTE REALITIES ON WHICH MR. A AND MR. B MUST AGREE.* [In this instance, Mr. A and Mr. B merely represent our own states of motion with the earth around the sun, as we travel with one velocity on a given day of the year, and another velocity six months later.] Physical experience shows that the earth's motion around the sun has no effect on the measured velocity of light in any direction: at all times of the year, and in all directions, this velocity is found to be 186,272 miles per second.

The absence of any effect of the earth's motion on any of our physical experiences was our main evidence (cf. page 38) for the truth of the General Law of Uniform Motion stated on page 36:

An observer in uniform motion finds all the same laws of physics to hold as does a stationary observer.

Since the earth's motion around the sun has no effect on the velocity of light either, we must simply conclude that the fact that light travels with a certain velocity is one of the laws of physics covered by this law of uniform motion. Thus, the General Law of

Uniform Motion should read:

An observer in uniform motion finds all the same laws of physics to hold as does a stationary observer, including the law that the velocity of light is 186,272 miles per second.

As we have seen on pages 62 and 63, this law, which correctly describes the facts of our experience, stands in flat contradiction with the common-sense view that time and distance are absolute realities on which all observers must agree.

[*Historical digression:* Before the experiment of Michelson and Morley was actually performed, everybody was half-consciously assuming that it would have the opposite outcome: that the apparent velocity of light *would* be affected by the motion of the person observing it, in just the way which common sense indicated. Thus, there seemed to be no sound reason to doubt the correctness of the traditional common-sense notions on time and space, or, for that matter, even to investigate these notions as carefully as we have in the preceding pages. But once the experiment had shown that even moving observers find the velocity 186,272 miles per second for light, it became possible for Einstein's genius to break through the stupefying influence of universally accepted tradition, to make the detailed analysis of the notions of time and space which we have repeated in the preceding pages, and to go on to the brilliant discoveries which we shall now explain.]

The common-sense view that time and distance are absolute realities thus stands in direct contradiction to the fact of experience discovered by Michelson. Opinion must yield to fact! The common-sense view is wrong; the answer which common sense gives to the three questions on page 58 is wrong; all the common-sense pictures and conclusions worked out on pages 58 through 63 are wrong. The true answer, whatever it may be, to the three crucial questions on page 58, must be compatible with the fact of experience that all uniformly moving observers find that light travels with the fixed speed

$$c = 186,272 \text{ miles per second}$$

in all directions.

What is the true answer to the three questions on page 58? As Einstein was the first to realize, *WE CAN FIND THIS OUT BY USING THE FACT THAT ALL UNIFORMLY MOVING OBSERVERS FIND THAT LIGHT TRAVELS WITH THE SPEED C IN ALL DIRECTIONS.*

Since this speed c is to play such an important role, it is very convenient to suppose that all our various uniformly moving observers choose their unit of distance equal to 186,272 miles, so that the speed of light is one unit of distance per second. *In order to have a convenient way of speaking, we will call this unit of distance, 186,272 miles, a "foot," so that the speed of light is 1 foot per second.* Even though our "foot" is a good deal longer than the conventional foot, this convenient manner of speech cannot cause us to make any mistake as long as we remember that "foot" really means 186,272 miles. In order to be conveniently able to describe the charts of events representing the notions of time and space belonging to our various observers, *we will agree that, from now on, each time we refer to one of these charts we will suppose that each observer draws his chart of events with a 1-inch separation between the horizontal lines representing events occurring 1 second apart, and with a 1-inch separation between the vertical lines representing events occurring 1 foot apart.* A uniformly moving observer finds that light from a flash travels to the right and to the left at 1 foot per second. Thus the observer's picture of the flash and the subsequent spread of light outward is as follows:

Fig. 44

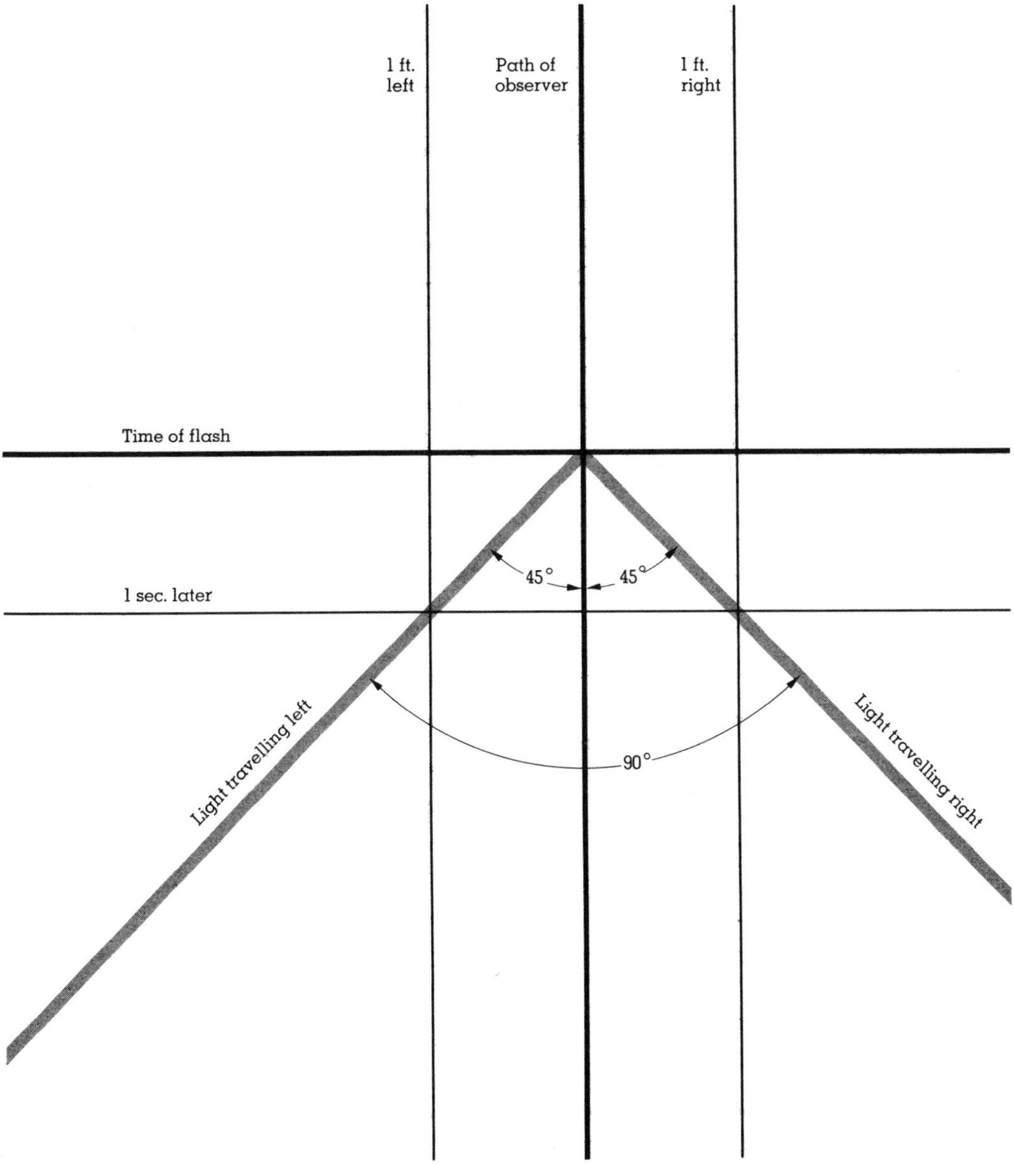

1 ft. left

Path of observer

1 ft. right

Time of flash

1 sec. later

45°   45°

90°

Light travelling left

Light travelling right

One second after the flash, as our observer reckons time, the leftward-traveling light is 1 foot to the right of the observer and the rightward-traveling light is 1 foot to the right of the observer. Since we have chosen our units so that light travels 1 foot per second, the lines in our chart which represent the successive positions of a packet of light always have a slope 1 in 1, or 45 degrees.

Let one uniformly moving observer, Mr. B, pass another uniformly moving observer, Mr. A. We shall look at things from Mr. A's point of view and suppose that Mr. A is stationary. At the instant Mr. A and Mr. B pass each other, they arrange to produce a flash of light. Mr. A is interested in determining what Mr. B's quantitative notions of time and space are. He already knows that Mr. B's notions of time and space must appear in his own chart in the general manner shown in Figure 41. He only needs to answer the three questions on page 58.

Suppose that in Mr. A's chart of events Mr. A draws two lines: the first representing all those events which, according to Mr. B, occur at the instant of the flash, the second representing all those events which, according to Mr. B, occur 1 second later; these are two lines whose exact slant and position we do not yet know but want to find out. The picture would look like this:

Fig. 45

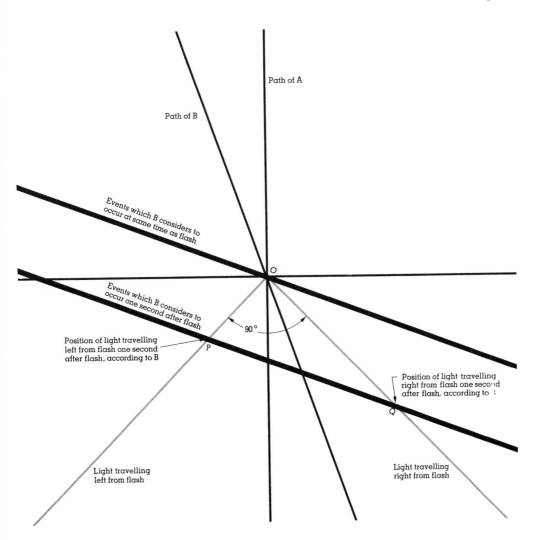

Path of A

Path of B

Events which B considers to occur at same time as flash

O

Events which B considers to occur one second after flash

Position of light travelling left from flash one second after flash, according to B

90°

P

Position of light travelling right from flash one second after flash, according to B

Q

Light travelling left from flash

Light travelling right from flash

The light gray lines represent the successive positions of the light traveling out from the flash, and consequently have 45-degree slopes. The dark gray straight line through O, representing those events which Mr. B considers to have occurred at the same time as the flash, is, of course, parallel to the dark gray line representing those events which Mr. B considers to have occurred 1 second later. The intersection of this latter dark gray line and the light gray line representing the successive positions of the rightward-traveling light from the flash represents the place, according to Mr. B, where the light arrives 1 second later, as Mr. B reckons time. Similarly, the intersection of the same dark gray line and the light gray line representing the successive positions of the leftward-traveling light from the flash represents the place, according to Mr. B, where the leftward-traveling light arrives 1 second later, as Mr. B reckons time.

Now, Mr. A knows that he and Mr. B agree that the light from the flash travels to the left and to the right at a speed of 1 foot per second. Hence he knows that the lines in Figure 41 representing all those events which Mr. B considers to be 1 foot to the left and to the right of himself, respectively, must respectively pass through the points P and Q. Thus, the preceding figure may be supplemented with additional information, like this:

Fig. 46

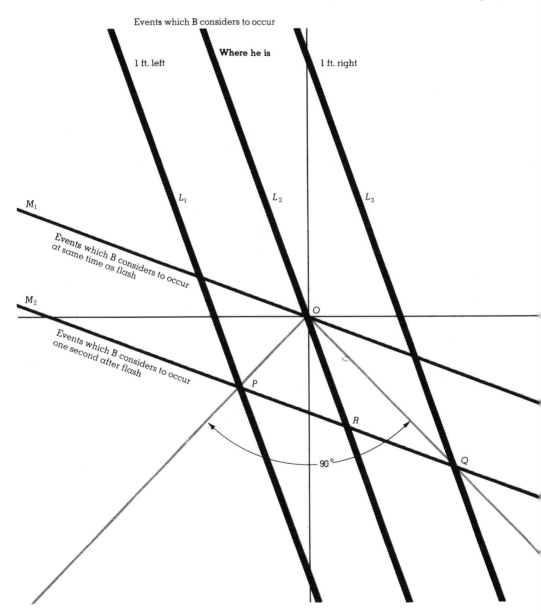

According to Mr. B, P is the point at which the left-ward-traveling light from the flash has arrived 1 second after the flash. Hence, according to Mr. B, the point P must be 1 foot to the left of his own position R 1 second later, and therefore the line L_1 parallel to his path L_2 must represent all events which occur 1 foot to the left of his position. Similarly, the line L_3 must represent all events which occur 1 foot to the right of his position.

We saw on pages 57–58 that Mr. A knows that the space between the lines L_1 and L_2 is equal to the space between the lines L_2 and L_3. We may consequently assert of the three points P, R, Q where the line M_2 crosses L_1, L_2, L_3 that the length in the chart from P to R is the same as that from R to Q, i.e., that R exactly bisects the segment PQ. Let us look in more detail at a part of Figure 46: the configuration formed by the lines L_2, M_2, and the two light gray lines representing the path of the light from the flash as it travels to the right and to the left. Turning the page a little so as to examine this configuration in a simpler position, we see that it looks like this:

Fig. 47

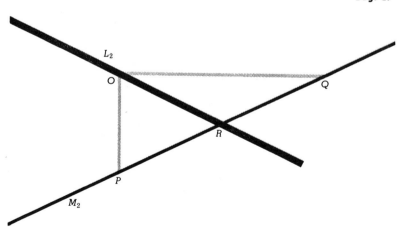

The line L_2 exactly bisects the segment PQ. The angle POQ is twice 45 degrees, or 90 degrees: a right angle. If we complete the half-rectangle formed by the two light gray lines, we see from the symmetry of the resulting figure that the line bisecting the diagonal PQ is just the line which passes through the fourth vertex of the rectangle:

Fig. 48

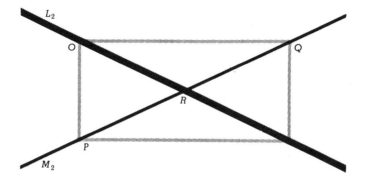

It is clear from the symmetry of this figure that certain angles must be equal to each other; of these, we mark the two most important:

Fig. 49

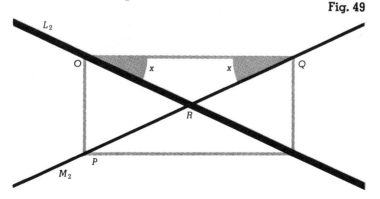

Now let us transfer this newly won information back to Figure 46. If we use the geometric fact that a line crossing two parallel lines makes equal alternate angles and that the pairs of opposite angles formed by the intersection of two lines are equal, we find many equal angles in that picture. Let us mark various important equal angles with symbols X, Y, etc.

Fig. 50

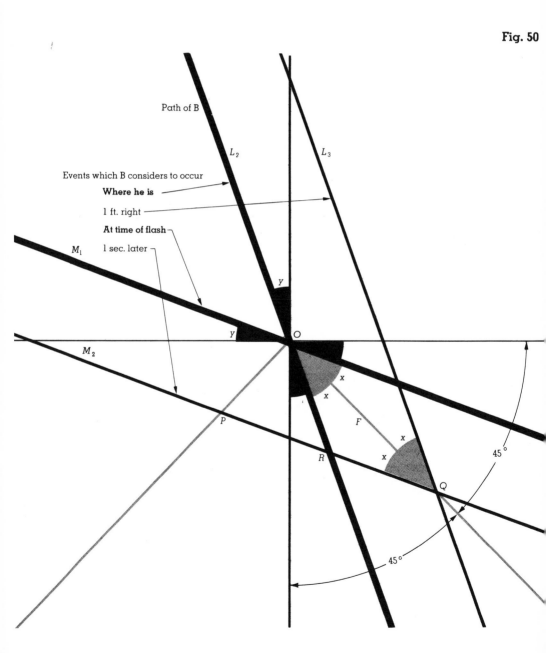

Path of B

L_2 L_3

Events which B considers to occur

Where he is

1 ft. right

At time of flash

M_1 1 sec. later

y

y O

M_2

x

x

P x

F

R x $45°$

Q

$45°$

It is clear from the equality of all the angles marked X that if we imagine a mirror reflection to be made in the light-brown line F, so that lines on opposite sides of F making equal angles with F are sent into each other, lines M_1 and L_2 are necessarily reflected into each other, and lines M_2 and L_3 are necessarily reflected into each other. Since such a reflection does not change any distances, the space in the chart between M_1 and M_2 must be equal to the space in the chart between L_2 and L_3.

Now we have seen that the parallel lines M_1 and M_2 represent all those events which Mr. B considers to occur at two given moments in time separated by an interval of 1 second, and that the parallel lines L_2 and L_3 represent all those events which Mr. B considers to occur at two given places in space separated by an interval of 1 foot.

So we have answered two out of three of the crucial questions on page 58!

From the equality of the two angles Y in the preceding chart, we have:

IN MR. A's CHART THE LINES WHICH REPRESENT ALL THOSE EVENTS WHICH MR. B CONSIDERS TO OCCUR AT THE SAME TIME MAKE THE SAME ANGLE WITH THE HORIZONTAL AS MR. B's PATH MAKES WITH THE VERTICAL.

From the fact that the space between M_1 and M_2 equals the space between L_2 and L_3, we have:

IN MR. A's CHART THE SPACE BETWEEN THE LINES REPRESENTING ALL THOSE EVENTS WHICH MR. B CONSIDERS TO OCCUR AT A GIVEN IN-STANT, ONE SECOND LATER, TWO SECONDS LATER, ETC. IS THE SAME AS THE SPACE BE-TWEEN THE LINES REPRESENTING ALL THOSE EVENTS WHICH MR. B CONSIDERS TO OCCUR AT A GIVEN PLACE, ONE FOOT TO THE RIGHT, TWO FEET TO THE RIGHT, ETC.

The only question that remains for Mr. A is: How big is this space? This question can be answered also; but it is better to pause and try to appreciate what we have already discovered than to go on and discover more.

The real facts about Mr. B's experience of time and distance, as we have just found them out, look like this in Mr. A's chart:

Fig. 51

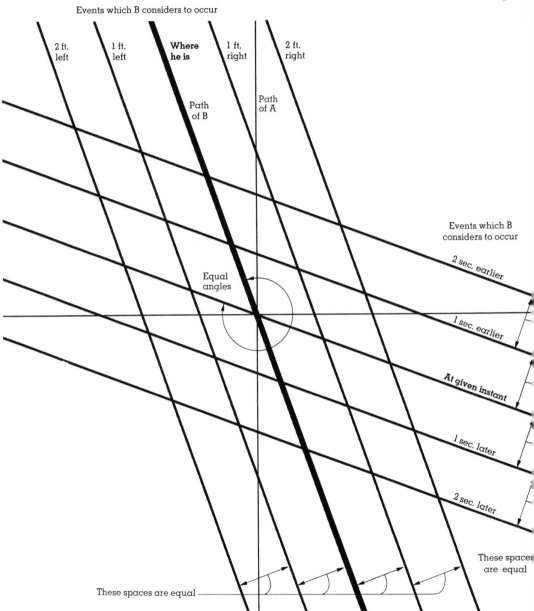

Events which B considers to occur

2 ft. left

1 ft. left

Where he is

1 ft. right

2 ft. right

Path of B

Path of A

Events which B considers to occur

2 sec. earlier

1 sec. earlier

At given instant

1 sec. later

2 sec. later

Equal angles

These spaces are equal

These spaces are equal

The point of view about time and space contained in the above picture (i.e., the point of view that the laws of physics are such that a moving observer, Mr. B, does not agree with a fixed observer, Mr. A, as to which sets of events occur at the same time, but instead has the notions of time and space indicated by the slanted lines in the diagram) is known as the *theory of relativity*. Einstein's discovery that this picture is correct, namely that two uniformly moving observers, each having a view of space and time as well justified as that of the other, disagree as to whether or not two given events occur at the same time; and hence that quantitative time is not an absolute property of the universe, is one of the most brilliant triumphs of the theoretical intellect. In the simple fact that the lines representing sets of events which Mr. B deems to have occurred at the same time are slanted we see vividly the overthrow of the classical notion of absolute time—the overthrow of a notion accepted for centuries without significant question by philosophers and scientists alike. As brilliant as the result is the crystalline simplicity of the Einsteinian reasoning which led us to it.

The name "relativity" comes from the fact that the basic deductions of the theory arise in a study of the motion of two observers, Mr. A and Mr. B, moving relatively to each other.

The mistaken common-sense view that distance and time are absolute objective realities on which Mr. A and Mr. B agree would cause Mr. A, instead of drawing the correct picture on the previous page, to draw the mistaken picture of the same facts given in Figure 42, which we repeat here to bring out its difference from the correct picture:

Fig. 52

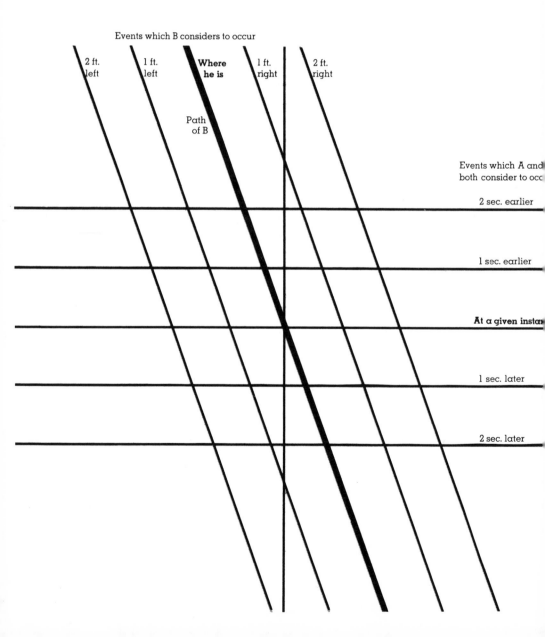

A striking difference indeed! How was it possible to miss this difference for so many centuries? How could everybody have held the mistaken view expressed in Figure 52 rather than the correct view expressed in Figure 51?

To answer this question we must remember that by "foot" we mean 186,272 miles. Therefore, even the earth's speed of 19 miles a second in its orbit around the sun is a very small fraction of the velocity of light: in the units which we have been using, only 1/9800 foot per second. Thus, if we draw the key lines of the picture in Figure 51 more in accordance with the general fact that we, as "Mr. A," rarely have to deal

with a particle or a Mr. B traveling more than a tiny
fraction of the speed of light, we see that it generally
looks more like this:

Fig. 53

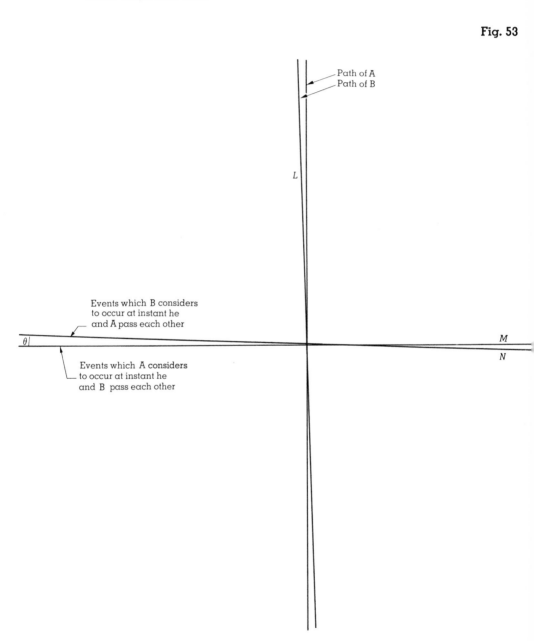

The line L in Mr. A's chart, representing Mr. B's position at successive instants of time, makes a small angle with the vertical. The line N, representing those events which, according to Mr. B, occur at the instant when he and Mr. A pass each other, makes an equal small angle with the horizontal.

The error in the common-sense view was to miss the small angle θ and imagine that the line N was really horizontal, and hence that it was the same as the horizontal line M.

But it is not perfectly horizontal; it is almost horizontal, and this only if Mr. B passes Mr. A with a speed equal only to a tiny fraction of the speed of light. If Mr. B passed Mr. A with a speed equal to a considerable fraction of the speed of light, the line N would be far from horizontal, and thus quite different from the horizontal line M. This shows that time is not an absolute objective reality which affects the physical experience of all observers in the same way and on which all observers agree. True, if Mr. B's speed relative to Mr. A is only a small fraction of the speed of light, Mr. A and Mr. B will differ only a little as to the time when a certain event happens. But, if Mr. B's speed relative to Mr. A is large, they can differ very considerably.

Let us look again at Mr. A's chart of space and time, with Mr. B's notions superimposed, as in Figure

51. Suppose that we draw the line L representing all those events which, *according to Mr. A,* occur 1 second *before* he and Mr. B pass each other, and also the line M representing all those events which, *according to Mr. B,* occur 1 second *after* he and Mr. A pass each other. Since these two lines are *not* parallel, they intersect in a certain point, E.

Fig. 54

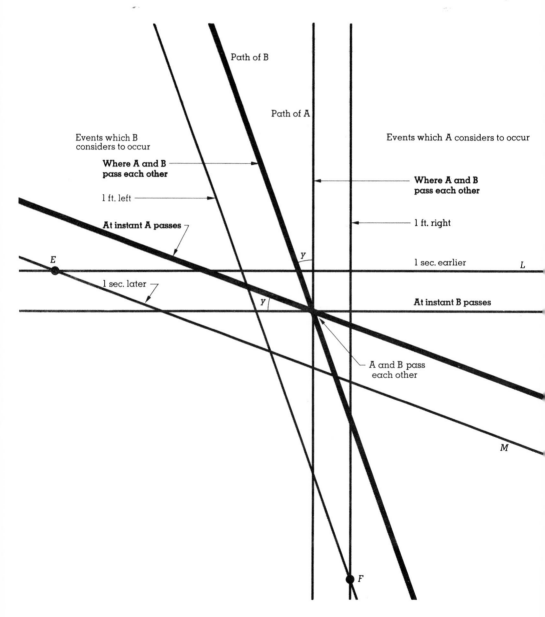

Path of B

Path of A

Events which B
considers to occur

Events which A considers to occur

**Where A and B
pass each other**

**Where A and B
pass each other**

1 ft. left

1 ft. right

At instant A passes

y

E

1 sec. earlier

L

1 sec. later

y

At instant B passes

A and B pass
each other

M

F

According to Mr. A, the event E happens 1 second *before* he and Mr. B pass each other; according to Mr. B, this event happens 1 second *after* Mr. A and Mr. B pass each other! This shows how little time really is an absolute objective reality and how much a uniformly moving observer's notion of time depends on his speed.

This is no more mysterious than the fact that Mr. A considers the event F to happen 1 foot to the right of the point in space where he and Mr. B pass each other, and Mr. B considers this event to happen 1 foot to the left of the point in space where Mr. A and Mr. B pass each other. Naturally, two observers moving relative to each other, each considering himself to be stationary, have different notions of what constitutes a fixed place in space. In exactly the same way they turn out to have different notions of what constitutes a fixed moment in time!

As appears vividly in the contrast between the correct picture in Figure 51 and the mistaken common-sense picture in Figure 52, time and space are on exactly the same footing in the correct picture. It is not the case, as the picture in Figure 52 would have it, that Mr. A and Mr. B disagree as to which events happen at the same place but agree on which events happen at the same time. They really disagree on both, having in consequence of their relative motion the systematic disagreement pictured in Figure 51. Now, it is perfectly comprehensible that, in consequence of their relative motion, various different uniformly moving observers all differ as to whether two events occurring at different times occur at the same place or not. Moreover, since all these different observers find that their physical experience is described by exactly the same laws of physics, so that there is no reason to say that one is any more right or wrong than another, it is perfectly clear that it does not make sense to ask whether two events occurring at different times "really" happen at the same place or not. From the point of view of some observers yes, from the point of view of other observers no. Thus the notion of an "objectively fixed point in space" cannot be defined except in a completely arbitrary way. Two different uniformly moving observers agree as much and as little as to whether two events occurring at different places "really" happen at the same time as they agree as to whether two events occurring at different times "really" happen at the same place. The notion of an "objectively fixed moment in time" cannot be defined except in a completely arbitrary way any more than can the notion of an "objectively fixed point in space."

That this seems perfectly all right in discussing space but surprising and disturbing in discussing time is just a consequence of the fact that most people have formed the habit of thinking in terms of the mistaken common-sense picture in Figure 52 instead of the correct picture in Figure 51. This is just a bad habit. The correct picture is no more illogical or complicated than the wrong one; indeed, since it treats time and space on the same footing, it is more symmetrical and hence more beautiful. After being in the bad habit of thinking in terms of the wrong picture for so long, the correct picture takes a little getting used to. But this is just a matter of practice.

To avoid error: It should not be thought that, because different uniformly moving observers consider different sets of events to take place at the same time or at the same place, time and space are mere subjective illusions, or that a moving observer has a distorted "subjective" impression of space and time. Nor should it be thought that, because we have frequently referred to "events which Mr. A considers to occur at the same time" or to "events which Mr. B considers to occur at the same time," we are implying that there is anything nonobjective about Mr. A's or Mr. B's point of view. Mr. A and Mr. B both arrive at their quantitative notions of space and time in the same way, each from a perfectly accurate and objective consideration of his own physical experience, in the manner explained on pages 4—7.

These notions are objective in the only sense in which any notion of science can be objective: in the sense that the notion is derived from the true facts of physical experience, and that the facts of physical experience are brought by the notion into harmonious order. Since Mr. A and Mr. B are moving relative to each other, they naturally come upon the events which constitute their common physical experience in a somewhat different manner; so that it is perfectly natural that the notions of time and space which Mr. A elaborates from his physical experience and which objectively govern Mr. A's experience should be somewhat different from the notions of time and space which objectively govern Mr. B's experience. Time and space are objective notions correctly describing each observer's physically lawful experience; but it is simply not the case that the statements "events E and F happen at the same place at different times" or "events E and F happen at different places at the same time" have an absolute meaning independent of the observer who makes them. Common sense, which perfectly well understands the way in which the truth or falsity of the second of these statements depends on the motion of the observer who makes the statement, must also learn to understand that the truth or falsity of the first of these statements depends equally on the motion of the observer who makes it! "Same time" is no more absolute than "same place."

This fact, which we have established on pages 68–75, may now be recognized to have some astounding consequences. Time and space are not absolute. What are absolute, in the sense of being the same for all observers, are the laws of physics: this is guaranteed by the General Law of Uniform Motion. Remember that for this law we have very powerful evidence: The motion of the earth in its yearly trip around the sun produces no seasonal variation in the laws of physics. The Michelson-Morley experiment provides a very exact confirmation of this last assertion. Now, we know that the notions of space and time for two uniformly moving observers are related as in the picture in Figure 51. This being the case, *THE GENERAL LAW OF UNIFORM MOTION TELLS US THAT ALL THE LAWS OF PHYSICS MUST BE SUCH AS TO APPEAR THE SAME TO TWO OBSERVERS WHOSE NOTIONS OF SPACE AND TIME ARE RELATED AS IN THE PICTURE IN FIGURE 51.*

Now, however, those laws of physics which are such as to appear the same to two observers whose notions of space and time are related as in the picture in Figure 51, naturally cannot each be exactly identical with the laws of physics which are such as to appear the same to two observers whose notions of space and time are related as in the common-sense picture in Figure 52. Thus we must expect that our new understanding of the manner in which space and time govern the physical experience of various uniformly moving observers should necessarily lead us to different laws of physics than those of Newton, who stated laws in accord with his view that time and distance are absolute realities.

Time and space, which are notions taken from physical experience, are themselves the foundation stones of physical law. Since we have seen that these foundations are different from what common sense would take them to be, we must expect that the structure of physical law built on these foundations is different from what was previously understood. This is the consideration which enabled Einstein to cap his revolutionary overthrow of the notion of absolute time with equally revolutionary discoveries in physics. Having followed him up to the discovery of the fundamental picture given in Figure 51, we shall now continue on the path laid out by him, and find, only a little farther on, some of his deservedly famous physical laws.

By applying the general principle emphasized on the previous page, we may deduce one of the most famous statements of relativity theory.

Let us imagine that, in virtue of some particular physical law, an observer, Mr. A, finds that a certain cause acting at a given time and place produces a certain effect in a distant place at some later time. Suppose that the effect is produced at a place so far distant, as Mr. A reckons distance, that the effect occurs before a ray of light starting out at the very time and place where the cause operates can travel to the place where the effect is produced. In Mr. A's chart of space and time the situation would look like this:

Fig. 55

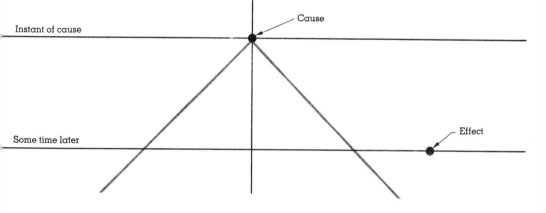

Mr. A may now ask himself: How would the situation appear to a second observer, Mr. B, moving with a speed somewhat less than but close to the speed of light? Using his knowledge of Mr. B's notions of time and space, as depicted in Figure 51, Mr. A could picture the answer as follows:

Fig. 56

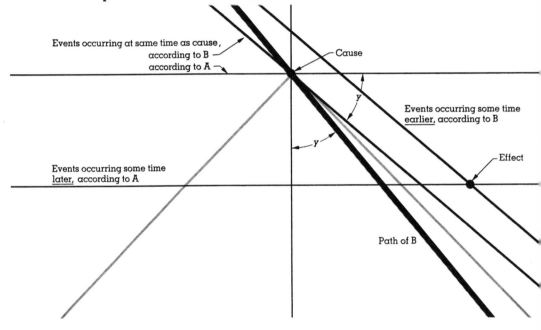

Events occurring at same time as cause, according to B

according to A

Cause

Events occurring some time earlier, according to B

Effect

Events occurring some time later, according to A

Path of B

Since Mr. B is traveling past Mr. A with a speed almost equal to the speed of light, his path in Mr. A's chart is a line inclined to the horizontal at only slightly *more* than 45 degrees. We know that the lines in Mr. A's chart representing those sets of events which Mr. B considers to occur at the same moment of time make the same angle with the vertical as Mr. B's path does with the horizontal. These lines are consequently inclined to the horizontal just a *little less* than 45 degrees. Since the point in Mr. A's chart representing the time and place of the effect is outside the angular region formed by the two 45-degree lines leading down out of the point representing the time and place of the cause, the line representing all those events which Mr. B takes to occur at the same time as the cause can well slant down *below* the point representing the effect.

Thus, from the point of view of Mr. B, the effect would take place before the cause! This is an impossible kind of physical law. But the physical law whose effects we have imagined Mr. A to witness would necessarily appear in just this impossible way to Mr. B. Hence such a law must be impossible. That is:

There can be no physical law according to which a cause acting at a given time and place could produce an effect at a distant place before a ray of light sent out from the time and place of the cause reached the place of the effect.

In particular,

No physical law by which a cause acting at a given place can have an effect at a distant place at the same instant is possible.

Now, according to Newton's law of gravitation, the gravitational force which one body exerts on another depends only on the distance between them; so that at the very instant that the first body is moved the second body experiences a changing gravitational force. But, we have just seen that this is impossible. Hence *NEWTON'S LAW OF GRAVITATION MUST BE WRONG*. When Einstein discovered this, he set out to find the correct law of gravitation, and in this way arrived at a theory of gravitation which he called the *general theory of relativity*. This theory is too complicated to be explained here, but it is interesting to know the reason why Einstein knew he should search for such a theory!

Another way in which the principle stated above is sometimes written is:

It is impossible by any means to produce an instantaneous effect at a distance.

These principles have been of great use to physicists, since they rule out many theories which might otherwise have had to be examined.

Another interesting example: If we had a perfectly rigid solid rod, then, at the very instant when we moved one of its ends, the other end would have to move also. But this would be an instantaneous effect at a distance. Hence *a perfectly rigid rod cannot possibly exist.*

This means that any actual rod will bend a little when one of its ends is moved, and that the bend in the rod will travel down the rod in the manner of a water wave, the other end moving only after the "bend" has had time to travel down the length of the rod. Similarly, gravitation must travel out from the body at which it originates in the manner of a wave.

It is worthwhile to exhibit graphically the general principle which is stated verbally on page 89. In terms of any observer's picture of space and time this may be done as follows:

Fig. 57

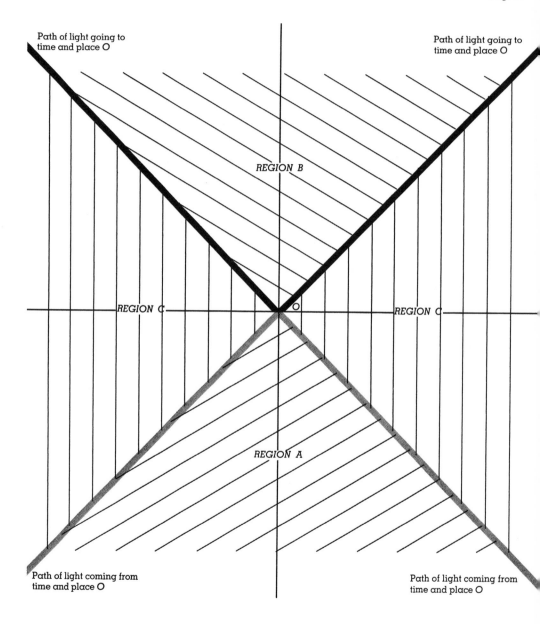

Path of light going to time and place O

Path of light going to time and place O

REGION B

REGION C

O

REGION C

REGION A

Path of light coming from time and place O

Path of light coming from time and place O

Straight lines drawn between O and a point in region C have a slope smaller than 45 degrees. Hence, if O were the cause of an effect taking place in the region C, the effect would take place before light from a flash at O reached the place of the effect. We have seen above that this is impossible. In this way the general principle on page 89 tells us that nothing that happens at the time and place marked O can possibly have any effect outside of region A. For exactly the same reason nothing that happens at a time and place lying outside region B can possibly have any effect on O. Things that happen at a time and place lying in region C can therefore neither affect nor be affected by anything that happens at the time and place marked O.

In particular, no physical particle or wave can travel from O to any point in region C. That is:

No physical particle or wave can travel with a speed greater than the speed of light.

The speed of light is consequently the greatest attainable speed.

This is one of Einstein's most famous conclusions.

It is not inappropriate to pause at this point to comment on two mistakes which people often make in talking about the theory of relativity. One is to imagine that the theory of relativity *assumes* that nothing can travel faster than light, and is built upon this assumption; as if, for instance, one could build an equally sensible theory by assuming that nothing can travel faster than sound. This is not so. We have not assumed but *proved* that nothing can travel faster than light, just as we have not assumed but proved that Newton's law of gravitation is wrong. All we have assumed is what is shown by the Michelson-Morley experiment: that all uniformly moving observers find that light travels with the same velocity in all directions. Why cannot the same thing be assumed about sound? Because the facts are different! A pilot traveling in a jet plane will find that the sound from an explosion behind him never reaches or passes him at all. In contrast, it follows from what has been said that a man in a rocket ship, no matter how fast he tries to go, will find in every case that light from a flash behind him catches up to him and passes him with a velocity that is always exactly the same. That there is something special about the velocity of light is not anything that Einstein made up: it is a fact of nature.

Another mistake is to imagine that the differences between the predictions of the theory of relativity and the predictions which follow from the conventional view that time and distance are the same for all observers are always small, just a few tiny fractions of a per cent. This is true only as long as we deal with particles and observers moving at a speed which is small compared with the speed of light, which is always the case in everyday experience. But if we deal with particles moving with a speed near the speed of light, the theory of relativity predicts effects which can become extremely large and quite unmistakable. For instance, one prediction of the theory of relativity is that a moving particle will become heavier the faster it goes. It can even be shown that as the speed of the particle approaches the speed of light more and more closely, the particle will become heavier without bound. (A somewhat sketchy account of this phenomenon will be found on page 116.) In the large atom smasher being built at Brookhaven, Long Island, which will accelerate particles to 99.98 per cent of the speed of light, the particles being accelerated will increase *fiftyfold* in weight.

It is no more possible for an engineer to build such an atom smasher in ignorance of the correct Einsteinian facts about time and space, as they are pictured in Figure 51, than it is possible for an automotive engineer to build a racing car to go 300 miles an hour if he thinks the right shape for a wheel is square. Nowadays Einstein's discoveries have become basic facts for certain kinds of engineering: every device in the design of which these discoveries play an important role is renewed proof of their correctness. The Michelson-Morley experiment merely happens to be the first piece of evidence for the truth of the theory of relativity which was discovered. Nowadays, innumerable pieces of evidence are known. It is by no means the case that the theory of relativity hangs on the thin thread of a belief that Michelson and Morley did not make an error of a few tenths of one per cent in their measurements.

A situation which might seem to contradict the law that nothing can travel faster than the speed of light is the following:

Suppose that Mr. B passes by Mr. A with a speed equal to three-fourths the speed of light at the same time that Mr. C passes by Mr. B in the same direction with a speed equal to three-fourths the speed of light. Does not Mr. C pass by Mr. A with the impossible speed of one and a half times the speed of light? This mistaken conclusion depends on the erroneous common-sense view of space and time. To see what the real situation is, we must think in terms of the correct picture, as it is shown on page 76. Let us work out the general result. Suppose that we have three uniformly moving observers, Mr. A, Mr. B, and Mr. C; that all three pass each other at the same moment; that according to Mr. A, who regards himself as stationary, Mr. B is traveling from left to right with a speed s; and that according to Mr. B, who also regards himself as stationary, Mr. C is traveling from left to right with a speed S. How do these two speeds "add up," i.e., how fast is Mr. C going from the point of view of Mr. A?

To answer this question, Mr. A must use a little high-school geometry.

Mr. A, who knows Mr. B's notion of space and time to be as they are represented on page 76, could picture all the facts of the situation in his chart of events as follows:

Fig. 58

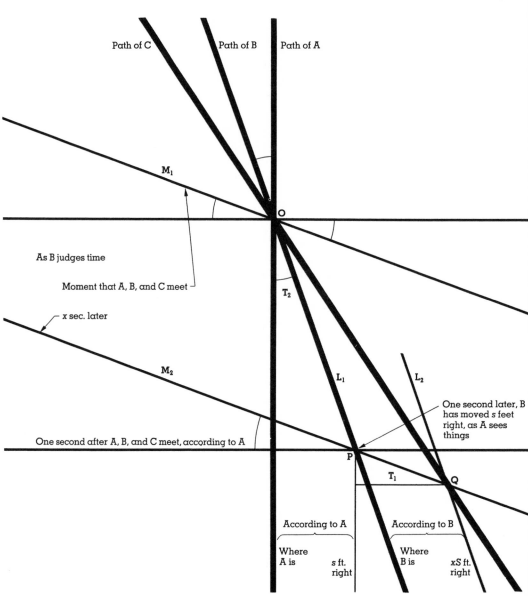

Path of C   Path of B   Path of A

M_1

As B judges time

Moment that A, B, and C meet

x sec. later

M_2

T_2

L_1 L_2

One second later, B has moved s feet right, as A sees things

One second after A, B, and C meet, according to A

P

T_1 Q

According to A   According to B

Where   s ft.   Where   xS ft.
A is   right   B is   right

Since Mr. B is traveling at a velocity of s relative to Mr. A, his path, the line L_1, has a slope of 1 in s. Moreover, since Mr. B considers Mr. C to be moving from left to right with speed S, then if Mr. B regards the events lying on lines M_1 and M_2 as being a certain number, x, seconds apart, Mr. B must consider that in this length of time Mr. C has moved x times S, or as it is written in algebra, xS feet to the right. Thus Mr. A knows that Mr. B must consider the events which lie on the lines L_1 and L_2 to be xS feet apart.

Now Mr. A knows the fact (stated on page 75) that what Mr. B considers to be a second of time and what Mr. B considers to be a foot of distance are represented by equal spaces in Mr. A's chart. Consequently, since Mr. B considers the lines M_1 and M_2 to represent events x seconds apart and considers the lines L_1 and L_2 to represent events xS feet apart, the gap in Mr. A's chart between the lines L_1 and L_2 must be exactly S times the gap between the lines M_1 and M_2. The lines M_1, M_2, and L_1, respectively, make the same angles with the vertical that the lines L_1, L_2, and M_2 make with the horizontal. Thus the line M_2 crosses the gap between the parallel lines L_1 and L_2 at exactly the same angle as the line L_1 crosses the gap between the parallel lines M_1 and M_2. Since the gap between L_1 and L_2 is exactly S times as large as the gap between M_1 and M_2, the segment PQ must be exactly S times as long as the segment OP. Now, the right triangles T_1 and T_2 have exactly equal smaller angles, so that they must be exactly similar in shape. Since the slanted side (hypotenuse) PQ of triangle T_1 is exactly S times as long as the slanted side (hypotenuse) OP of triangle T_2, it must be true that T_1 and T_2 are exactly in the proportion of S to 1.

The long perpendicular side of triangle T_2 represents 1 second of time in Mr. A's chart, while its short perpendicular side represents a distance of s feet. Since we have agreed (cf. page 66) that Mr. A is to represent a distance of 1 foot and a time interval of 1 second in his own chart by a space of 1 inch, the long perpendicular side of triangle T_2 is 1 inch long, while its short perpendicular side is s inches long. We have just seen that triangle T_1 is exactly similar to T_2 in shape but reduced in the proportion of S to 1. Thus the long perpendicular side of triangle T_1 is S inches, while its short perpendicular side is S times s, or sS inches. The configuration of the two triangles consequently is as follows:

Fig. 59

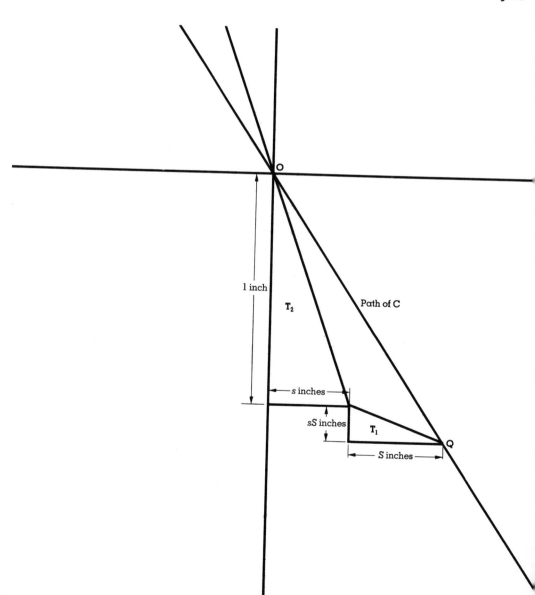

It is then clear that the point Q is $1 + sS$ inches
down and $s + S$ inches over from the point O. That
is, Q represents an event which occurs, according to
Mr. A, $1 + sS$ seconds after the event O, i.e., $1 + sS$
seconds after Mr. A, Mr. B, and Mr. C all pass each
other; and $s + S$ feet to the right of the position in
space where Mr. A, Mr. B, and Mr. C pass each other.
Hence, as far as Mr. A is concerned, Mr. C has moved
$s + S$ feet to the right in $1 + sS$ seconds. Since speed
is the quotient of distance moved by time it took to
move this distance, Mr. C's speed, as Mr. A sees it,
is simply

$$\frac{s + S}{1 + sS}$$

That is:

> If a uniformly moving observer, Mr. A, finds a
> second uniformly moving observer, Mr. B, to be
> traveling at a speed equal to a certain fraction s
> of the speed of light, and Mr. B finds a certain
> particle C to be moving at a fraction S of the
> speed of light, then Mr. A finds the particle C to
> be moving with a speed equal to
>
> $$\frac{s + S}{1 + sS}$$
>
> of the speed of light.

Our common-sense feeling for the addition of velocities would tell us to forget the denominator, and take the sum $s + S$ of the two speeds for the speed of Mr. C as Mr. A sees it. But this is merely another mistake, a consequence of the basic mistake of believing that Mr. A and Mr. B agree on time and distance.

According to the formula which we have just discovered, if Mr. B goes past Mr. A at 3/4 the speed of light, and Mr. C goes past Mr. B at 3/4 the speed of light, then Mr. C goes past Mr. A at

$$\frac{\frac{3}{4} + \frac{3}{4}}{1 + \frac{3}{4} \cdot \frac{3}{4}} = \frac{\frac{3}{2}}{\frac{25}{16}} = \frac{3}{2} \cdot \frac{16}{25} = \frac{24}{25} = 96 \text{ per cent}$$

of the speed of light. It appears after all that we did not have to worry about two speeds each equal to 3/4 the speed of light adding up to a speed greater than that of light! Indeed, even if Mr. B goes past Mr. A at 9/10 the speed of light, and Mr. C goes past Mr. B at 9/10 the speed of light, then according to this same formula Mr. C goes past Mr. A at just

$$\frac{\frac{9}{10} + \frac{9}{10}}{1 + \frac{9}{10} \cdot \frac{9}{10}} = \frac{\frac{18}{10}}{\frac{181}{100}} = \frac{180}{181}$$

of the speed of light.

If one of the particles B which the Brookhaven, Long Island, atom smasher accelerates to 99.98 per cent of the speed of light carries along with it an equally powerful atom smasher which is used to accelerate other particles C to 99.98 per cent of the speed of light, then an observer sitting on Long Island would find the particles C traveling at

$$\frac{.9998 + .9998}{1 + (.9998)(.9998)} = .99999996$$

of the speed of light.

All this provides vivid confirmation of our earlier assertion that it is impossible for any particle to attain the speed of light: even after the speed of light has "almost" been reached, the remaining "tiny increase" in speed is just as hard to attain as was the full speed of light in the first place. It is also amply plain that for speeds near the speed of light the differences between the relativistic truth and the common-sense view become quite considerable.

How great is the difference between the relativistic truth and the common-sense view in an everyday situation? Suppose, for instance, that two automobiles pass a lamppost on a superhighway in opposite directions, each at 60 miles per hour. How fast do the two cars pass each other? Now, 60 miles per hour is 1 mile per minute, or 1/60 mile per second; that is

$$\frac{1}{(60)\,(186,272)}$$

of the speed of light. The two cars consequently pass each other at

$$\frac{\dfrac{1}{(60)\,(186,272)} + \dfrac{1}{(60)\,(186,272)}}{1 + \dfrac{1}{(60)\,(60)\,(186,272)\,(186,272)}}$$

of the speed of light, i.e.,

$$\frac{120}{1 + (60)\,(60)\,(186,272)\,(186,272)} = 119.9999999999999$$

miles per hour. In this case the common-sense answer of "120 miles per hour" is exceedingly close to the truth.

Of two uniformly moving observers, each considers himself to be stationary and the other to be moving. To this extent the relation between the two observers is perfectly symmetrical. For the further progress of our investigation, it is important to show that the relation between the two observers is perfectly symmetrical in all respects. We now turn our attention to this task. Let us consider our two uniformly moving observers, Mr. A and Mr. B; let us suppose that from Mr. A's point of view, Mr. B is traveling from left to right at S feet per second. Then, from Mr. B's point of

view, how fast is Mr. A traveling? Mr. A can figure
this out, using the picture on page 76, by marking and
studying the lines indicated in the following figure
(*M₁* and *L₂* are drawn through the point *P*):

Fig. 60

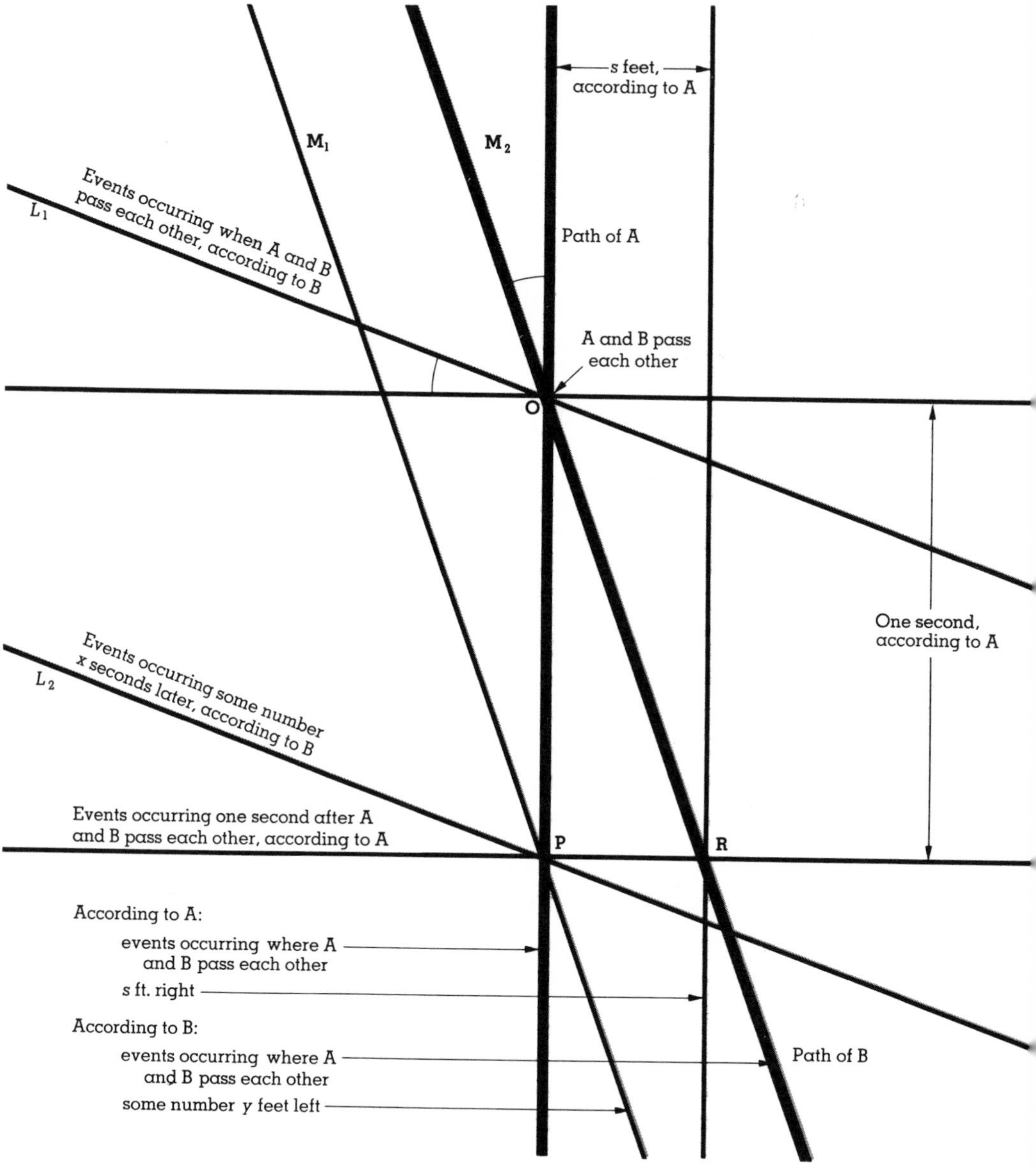

M_1

M_2

L_1 Events occurring when A and B
pass each other, according to B

s feet,
according to A

Path of A

A and B pass
each other

O

One second,
according to A

L_2 Events occurring some number
x seconds later, according to B

Events occurring one second after A
and B pass each other, according to A

P   R

According to A:

events occurring where A
and B pass each other

s ft. right

According to B:

events occurring where A
and B pass each other

some number y feet left

Path of B

The parallel lines L_1 and L_2 make the same angle with the vertical as the parallel lines M_1 and M_2 make with the horizontal. Hence the segment OP crosses the gap between L_1 and L_2 at the same angle as the segment PR crosses the gap between M_1 and M_2. Since the segment PR is exactly S times as long as the segment OP, the gap between M_1 and M_2 must be exactly S times as long as the gap between L_1 and L_2. Now, Mr. A knows that what Mr. B considers to be a second of time and what Mr. B considers to be a foot of distance are represented by equal spaces in Mr. A's chart. Since the gap between M_1 and M_2 represents y feet for Mr. B, and the gap between L_1 and L_2 represents x seconds for Mr. B, and since the gap between M_1 and M_2 is S times the gap between L_1 and L_2, it follows that y and x are exactly in the proportion of S to 1. Thus, Mr. B considers that when Mr. A arrives at the point P, the time is x seconds later, and Mr. A has moved Sx feet to the left. Since speed is simply the quotient of distance traveled by time it took to travel that distance, Mr. B considers that Mr. A is traveling from right to left with a speed of S feet per second. Thus

If the first of two uniformly moving observers considers the second to be moving from left to right at a certain speed S, then the second observer considers the first to be moving from right to left with the same speed S.

In this particular case, common-sense feeling turns out to be perfectly correct. That this must be the case can also be shown in another way. Suppose that we had two uniformly moving observers, Mr. A and Mr. B; and that from Mr. A's point of view Mr. B was moving with a speed S, while from Mr. B's point of view Mr. A was moving with a smaller speed s. Then Mr. A could use this discrepancy to try to convince Mr. B that Mr. A was really standing still and that Mr. B was really moving. Since Mr. B would have to admit that Mr. A had a smaller velocity than that which Mr. A ascribed to Mr. B, he would have to admit that there was some truth in Mr. A's claim.

Since in virtue of the General Law of Uniform Motion a uniformly moving observer can have no indication that he is not stationary, we are forced to that conclusion that $s = S$. This is the same conclusion that was proved in a different way on the preceding page.

Even more: The relation between the two uniformly moving observers Mr. A and Mr. B must be entirely symmetrical, since any asymmetry could be used to buttress the claim of one or the other to be the one who was really stationary. Thus, any peculiarity in Mr. B's notions of time and space which Mr. A observes must correspond exactly to a similar peculiarity in Mr. A's point of view which Mr. B observes. Each must direct the same "accusations" at the other!

In Mr. A's chart of events Mr. B's notions of space and time appear as follows:

Fig. 61

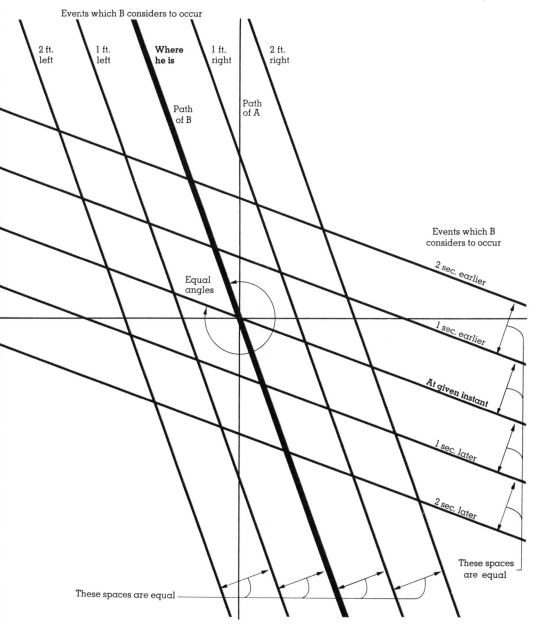

This is just the picture in Figure 51.

In Mr. B's chart of events Mr. A's notions of space
and time appear as follows:

Fig. 62

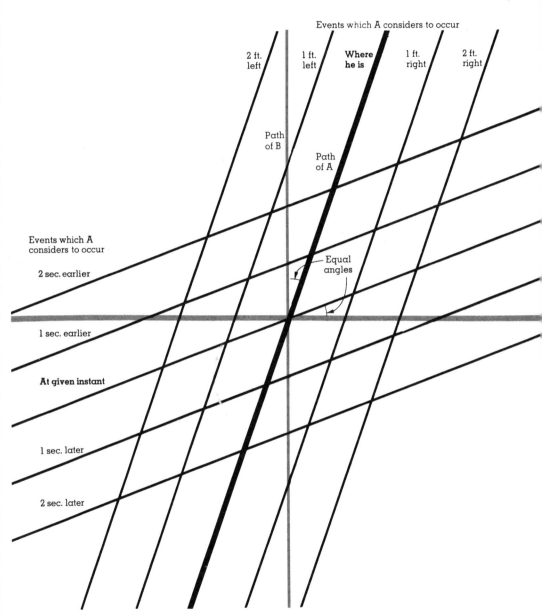

The angles between the dark gray and the black lines are the same in the preceding diagram as in the diagram on page 105. The slope of these lines represents, in one case, Mr. B's velocity as judged by Mr. A, and, in the other case, Mr. A's velocity as judged by Mr. B. We know that the velocity which Mr. B ascribes to Mr. A, Mr. A ascribes to Mr. B.

The relationship between Mr. A and Mr. B is perfectly symmetrical, so that the spaces between the slanted lines in the picture on the preceding page should be exactly as large as the spaces between the slanted lines in the picture in Figure 61.

How large are these spaces? Of the three questions on page 58, this is the only one which we have not answered. The question may be asked as follows: How long a time does Mr. B consider to have elapsed between the instant when he passes Mr. A and the instant when he reaches that point in his path which, *according to Mr. A's notion of time,* he reaches after 1 second?

We shall now answer this question.

We are given two uniformly moving observers, Mr. A and Mr. B. Mr. A considers Mr. B to be traveling at a certain fraction S of the speed of light. Our question is: How long a time does Mr. B consider to have elapsed between the instant when he passes Mr. A and the instant when he reaches that point in his path which, according to Mr. A's notion of time, he reaches after 1 second? Suppose that Mr. B considers this length of time to be a certain number, x, of seconds. Since Mr. B and Mr. A are related in a perfectly symmetrical way, it follows that the length of time which Mr. A considers to have elapsed between the instant when he passes Mr. B and the instant when he reaches that point in his path which, according to Mr. B's notion of time, he reaches after 1 second is also x seconds. Thus, the amount of time which Mr. A considers to have elapsed between the instant he passes Mr. B and the instant when he reaches that point in his path which, according to Mr. B's notion of time, he reaches after x seconds bears the same proportion to x as x does to 1, that is, is x times x, or x^2 seconds. In Mr. A's chart all this would be pictured as follows:

Fig. 63

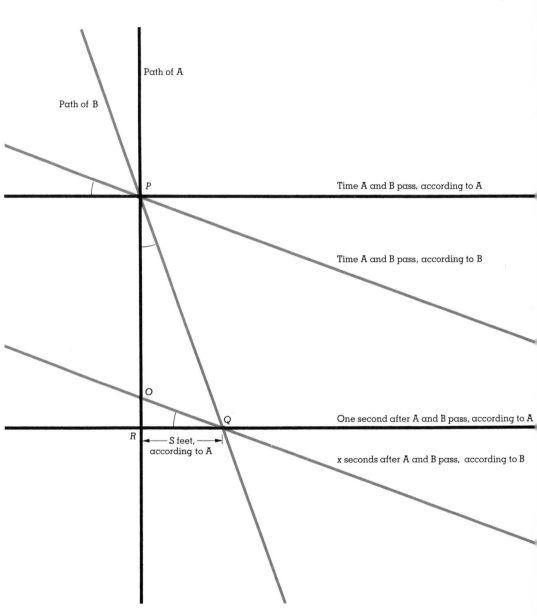

Path of A

Path of B

P

Time A and B pass, according to A

Time A and B pass, according to B

O

Q

One second after A and B pass, according to A

R

$\longleftarrow$ S feet, $\longrightarrow$
according to A

x seconds after A and B pass, according to B

Note that the point Q represents an event which Mr. A considers to occur 1 second after the time when Mr. A and Mr. B pass; hence, as stated above, the point Q represents an event which Mr. B considers to occur x seconds after Mr. A and Mr. B pass. The points on the dark gray line through Q, making an angle with the horizontal equal to that which the path of B makes with the vertical, consequently represent all these events which Mr. B considers to occur x seconds after Mr. A and Mr. B pass. The intersection O of this line with the path of Mr. A is therefore the point which, according to Mr. B, Mr. A reaches x seconds after Mr. A and Mr. B pass. By what has been said above, Mr. A considers that he reaches this point x^2 seconds after Mr. A and Mr. B pass. The segments PR and RQ represent 1 second of time and S feet of distance, respectively, as Mr. A reckons time and distance. Consequently, these segments are 1 inch and S inches, respectively (cf. page 66). By what we have said just above, PQ represents x seconds of time, as Mr. B reckons time, and PO represents x^2 seconds of time, as Mr. A reckons time. Since we are looking at Mr. A's chart, the segment PO is x^2 inches long. Now, the two right triangles PRQ and QRO have the same smallest angle, and must be exactly proportional to each other. Since the long and the short perpendicular sides PR and RQ of triangle PRQ are 1 inch and S inches respectively, the short side OR of triangle QRO must likewise be S times the long side RQ of this triangle. But RQ is S inches. Thus OR must be S times S, or S^2, inches. Since PO is x^2 inches, OR is S^2 inches, and the whole segment PR is 1 inch, it follows that $x^2 + S^2 = 1$; thus $x^2 = 1 - S^2$, so that taking the square root of both sides of this equation we find that $x = \sqrt{1 - S^2}$.

That is: Mr. A considers P and O to be events occurring $1 - S^2$ seconds apart; Mr. A considers P and R to be events occurring 1 second apart; and Mr. B considers P and Q to be events occurring $\sqrt{1 - S^2}$ seconds apart.

Stating this more carefully, we find that:

If Mr. A, the first of two uniformly moving observers, finds a second observer, Mr. B, to be

traveling past him at a fraction S of the speed of light, Mr. A will find that Mr. B considers that only $\sqrt{1-S^2}$ seconds of time have elapsed when Mr. B reaches the point in his path which Mr. A considers him to reach after 1 second, and Mr. B will have an exactly corresponding opinion about Mr. A.

Now, Mr. B forms his quantitative notions of time by observing the physical processes about him, just as Mr. A does. Moreover, by virtue of the General Law of Uniform Motion, Mr. B finds no abnormalities or disproportions in the rate at which various physical processes take place in his vicinity. Thus, what Mr. A considers to be Mr. B's mistaken impression of the rate at which time passes must be a consequence of the fact that all the physical processes in Mr. B's environment are slowed up, from Mr. A's point of view, in the proportion of $\sqrt{1-S^2}$ to 1.

From the point of view of a given uniformly moving observer (who naturally considers himself to be stationary) all physical processes occurring in a physical system traveling past him at a fraction S of the speed of light will be slowed from their ordinary rates in the proportion of $\sqrt{1-S^2}$ to 1.

SO THAT MOVING CLOCKS OF EVERY DESCRIPTION RUN SLOW!

This is another of the most famous of Einstein's laws.

Time and space occur in the pictures on pages 105 and 106 in an exactly symmetrical way. Now, we saw on page 109 that if the segment PR represented a second of time, as Mr. A reckons time, then the segment PO represents $1-S^2$ seconds of time, as Mr. A reckons time, and the segment PQ represents $\sqrt{1-S^2}$ seconds of time, as Mr. B reckons time. Using the symmetry which we have noted, we may conclude at once that if the segment $P'R'$ in the following picture represents a foot of distance, as Mr. A reckons distance, then the segment $P'O'$ represents $1-S^2$ feet of distance, as Mr. A reckons distance, and the segment $P'Q'$ represents $\sqrt{1-S^2}$ feet of distance, as Mr. B reckons distance.

On the other hand, the line L would represent the successive positions of a particle traveling at the same speed as Mr. B, but to the right of him. Since $P'O'$ represents $1 - S^2$ feet for Mr. A, Mr. A would say that this particle is $1 - S^2$ feet right of Mr. B. Since $P'Q'$ represents $\sqrt{1 - S^2}$ feet for Mr. B, Mr. B would say that this particle is $\sqrt{1 - S^2}$ feet right of himself. Now, the ratio between y^2 and y is y^2/y, or y; which is the same as the ratio of y to 1. Thus the

Fig. 64

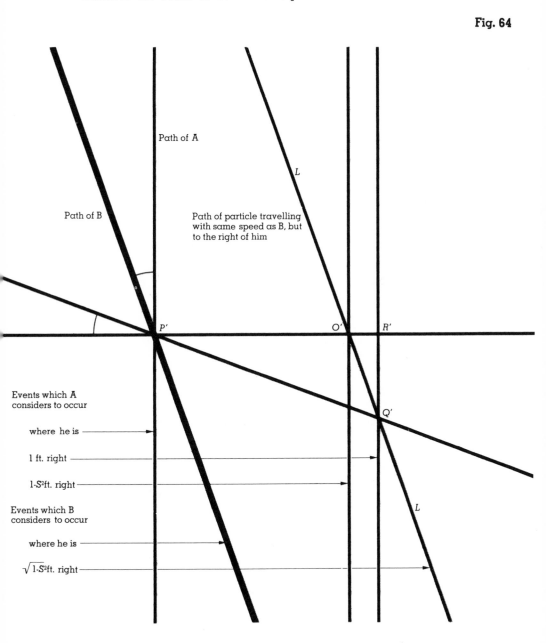

Path of A

L

Path of B

Path of particle travelling
with same speed as B, but
to the right of him

P'　　　O'　R'

Events which A
considers to occur

where he is

1 ft. right

1-S²ft. right

Q'

Events which B
considers to occur

L

where he is

$\sqrt{1\text{-}S^2}$ft. right

ratio between $1 - S^2$ and $\sqrt{1 - S^2}$ is just that between $\sqrt{1 - S^2}$ and 1. We consequently find from the above figure that:

If Mr. A, the first of two uniformly moving observers, finds a second observer, Mr. B, to be traveling at a fraction S of the speed of light, and if Mr. A also finds a particle to be traveling at the same speed as Mr. B but $\sqrt{1 - S^2}$ feet to the right of Mr. B, Mr. A will also find that Mr. B considers the particle to be one foot to the right of him.

Thus, just as in the case of time, as we discussed it on page 110, this must mean that:

From the point of view of a given uniformly moving observer (who naturally considers himself to be stationary) all the lengths in a physical system traveling at a fraction S of the speed of light will be shrunk from their ordinary size in the proportion of $\sqrt{1 - S^2}$ to 1.

So that moving rulers of every kind shrink!

The last principle stated is only correct in the form in which it is given when we deal with only a single dimension of space instead of the three dimensions which space actually has. If we took the trouble to figure out the minor corrections which this fact requires us to make in all that has been said till now, we would find that the "ruler-shrinkage" just referred to takes place only in the direction of motion, and that lengths in perpendicular directions are left unaffected.

The four facts that have just been emphasized are among the most famous discoveries of Einstein. To appreciate them a little better, let us see how they apply to various situations.

If a jet plane travels 2500 miles per hour, how much slowing of time and shrinkage of length aboard the jet will we observe watching from outside? This is easy to compute. A speed of 2500 miles per hour is

$$\frac{2500}{60 \times 60} = \frac{25}{36} \text{ miles per second}$$

or $\dfrac{25}{(36)(186,272)}$ the speed of light

This is about .000004 of the speed of light. So

$$\sqrt{1 - S^2} = \sqrt{1 - (.000004)(.000004)}$$
$$= .999999999992$$

and time runs slower for the pilot of the jet plane by 8 parts in a trillion. This is a loss of 2 seconds every 10,000 years—which shows why the effect in question was missed for so long.

On the other hand, if the speed S of a moving physical system is close to the speed of light, so that S, and hence S^2, is just a little less than 1, then $1 - S^2$ and hence $\sqrt{1 - S^2}$ is practically zero. Thus, the rate of physical process, and the length of physical objects, in a system traveling at a speed close to that of light, both appear to a "stationary" observer to have shrunk nearly to zero.

A number of very interesting questions are connected with the formulae which we have found. In the first place, it should be noted that, since S^2 is always positive if the speed S is different from zero, the factor $\sqrt{1 - S^2}$ is always less than 1 if S is different from zero. Consequently, the rate at which physical processes occur on a moving system is, from the standpoint of a stationary observer, always *less* than their ordinary rate. This remark leads to the following

PARADOX OF THE ADVENTUROUS TWIN:

Of the twins Mr. Y and Mr. Z, Mr. Y is adventurous while Mr. Z is stodgy. When they are both twenty, Mr. Z consequently decides to remain at home on earth, while Mr. Y travels to distant points in a fast rocket ship which attains velocities close to that of light for long periods. Thus, from Mr. Z's point of view, Mr. Y ages at a diminished rate, and sometimes at a greatly diminished rate. When Mr. Z is sixty years old, Mr. Y returns. But, since from Mr. Z's point of view Mr. Y has been aging at a diminished rate, Mr. Y is only thirty years old when he returns.

What is the paradox? It is this: Since from Mr. Y's point of view he is the one who is stationary, while Mr. Z is moving, then from Mr. Y's point of view Mr. Z is aging at a diminished rate. Consequently, if Mr. Y is thirty years old when he returns, Mr. Z must be even younger.

This apparent paradox has troubled many students of relativity and has even been used by opponents of relativity in efforts to support the claim that the theory of relativity is self-contradictory. But, if we examine the conditions of the apparent paradox more carefully, we shall see that the apparent inconsistency rests on a misconception.

What is wrong? The fact is that Mr. Y and Mr. Z are not symmetrically related in this problem. Mr. Z is not accelerating at any time. Mr. Y is accelerating for at least part of the trip. While a uniformly moving observer observes no physical effects which might indicate to him that he is not at rest, a nonuniformly moving observer observes all sorts of physical effects which prove conclusively to him that his state of motion is changing. Consider once more the different experiences which one has in a car moving along steadily and in a gradually accelerating car. A passenger in the latter feels himself pushed back into the seat. Consider finally the very different experience of a man whose car hits a concrete wall! He would not be tempted to claim either that his state of motion did not change during the crash or that he observed no exceptional effects during the crash. Thus Mr. Y and Mr. Z are not symmetrically related, and there is no paradox: Mr. Y really does return thirty years old when Mr. Z is sixty. A paradox could arise only if neither Mr. Y nor Mr. Z accelerated at all, so that they were symmetrically related. But in this case, Mr. Y could never turn around and come back once he started out on his trip, so Mr. Y and Mr. Z would never again meet after the time when they parted. So much for the apparent inconsistency.

Does the fact that when Mr. Y returns he is younger than Mr. Z not prove that it is better to travel than not to travel, since by traveling Mr. Y causes time to pass more slowly for him than it does for Mr. Z, and hence stays young while Mr. Z ages? Alas, no. Since all the physical processes which Mr. Y has observed aboard his rocket ship while on his trip have moved through a succession of stages corresponding exactly to the ten years which he has aged, Mr. Y has been able to enjoy only ten years of life while Mr. Z has enjoyed forty. What is given unto both Mr. Y and Mr. Z is a life span encompassing threescore and ten years of experience. The relative rate at which they use up this common life span is not defined. This is the fact: that it appears paradoxical is only another result of the mistaken habit of considering time to be absolute.

It is true, of course, that by traveling Mr. Y has the option of postponing his span of experience to a later epoch of world history. If he thinks that the future will be better than the age in which he finds himself, he may prefer to do this. On the other hand, if, like George Orwell, he thinks that the future will be worse, he will prefer to live out his life in the present. So then Mr. Y does have an option. On the other hand he has this option quite irrespective of the theory of relativity. If he wants to live in the future, he has only to have himself frozen in a suitable, careful way, and later thawed out and revived.

In the preceding pages an account of Einstein's discoveries about the basic notions of time and space has been presented. It has also been remarked that these notions, time and space, are the foundation stones of physical law. Since we have seen that these foundations are different from what common sense took them to be, we must expect that the structure of physical law built on these foundations is different from what has been traditionally understood. Thus Einstein's discoveries about time and space imply a new physics. This new physics, some of whose laws we have discussed on pages 89–93, was worked out in considerable detail by Einstein. To continue this book would be to exhibit these new physical laws and to follow Einstein's derivations of them. Unfortunately, such an undertaking is too ambitious for the present little book for the following reason: Even the simplest of these new laws of physics could be stated only in terms of the notions of physics: energy, mass, momentum. To explain the meaning of these concepts in a manner sufficiently careful and precise to make possible a convincing derivation of Einstein's physical laws would be to retrace the whole of the early history of physics. To attempt such a task would be to enlarge the present book beyond all reasonable length. The reader who wishes to study these very interesting questions is best advised to study the ordinary college physics first, so as to become familiar with the meaning and significance of mass, energy, momentum, etc. This done, the Einsteinian modifications of Newton's laws of physics can be more easily presented and more readily grasped.

Nevertheless: Just to indicate, in a qualitative way, what Einstein's new physical laws are like, we shall touch upon one of the effects which he discovered.

We have seen that no particle can ever be made to travel faster than the speed of light. This must mean that as the speed of a particle approaches the speed of light, it becomes more and more difficult to bring about increments in its speed. That is, the effect of a fixed force of acceleration exerted on a particle must become smaller and smaller as the velocity of the particle approaches the velocity of light. Now, the proportionate resistance which a body offers to acceleration is called its *mass*. Thus we are led to the conclusion:

The mass of a moving body must increase as its velocity increases, and increase without limit as its velocity nears the velocity of light.

Einstein was able to find the precise quantitative expression for this increase in mass of a moving body; but, for the reasons just explained, I shall not attempt to exhibit the details of his arguments on this score.

If energy, say in the form of heat, is communicated to a body, then the individual molecular particles constituting the body are caused to move more rapidly. By the principle just stated, each of these particles, and hence the body as a whole, must become more massive. Thus:

An increase in the energy of a body must be connected with an increase in its mass.

Einstein was able to show that the increase in mass must be exactly proportional to the increase in energy, and to show that the correct factor of proportionality is the square of the velocity of light. This conclusion, expressed in the famous formula

$$\text{Energy} = \text{Mass} \times (\text{velocity of light})^2$$

revealed, in 1905, the tremendous energies locked up in the atomic nucleus.

EPILOGUE

A number of books are available for the reader who would like to pursue the topics in relativistic mechanics not discussed in the present book. A deduction of the quantitative laws governing the increase of mass with velocity and the relation between mass and energy is given in Chapter VI of Herbert Dingle's little book *The Special Theory of Relativity*. This topic is also treated in *The Einstein Theory of Relativity* by Lillian R. Lieber and Hugh Gray Lieber, which is a book of intermediate difficulty written in an unusual style. The Liebers' book also contains an account of the "general theory of relativity," that is, the relativistic theory of gravitation. Rather elementary qualitative discussions of relativity theory, including the general theory, are given in Einstein's *Relativity—the Special and General Theory*, and in Part III of *The Evolution of Physics* by Einstein and Leopold Infeld. This latter book is especially valuable in that it discusses a number of strategic topics in the general history of physics, thereby putting the theory of relativity into a broader perspective. There exists a large number of treatments of the theory of relativity which are "popular" in the sense of being discursive or anecdotal. Two of the best of these are *From Copernicus to Einstein* by Hans Reichenbach, and *The Universe and Dr. Einstein* by Lincoln Barnett, which Mentor Books has published in an inexpensive paperback edition.

One of the best of the systematic advanced mathematical treatments of the theory of relativity is *The Classical Theory of Fields* by L. D. Landau and E. Lifschitz. This work also treats the gravitational theory.